Fatigue of Composite Materials

FATIGUE OF COMPOSITE MATERIALS

Ramesh Talreja, Ph.D., D.Sc.
The Technical University of Denmark

TECHNOMIC
PUBLISHING CO., INC.

LANCASTER • BASEL

Published in the Western Hemisphere by
Technomic Publishing Company, Inc.
851 New Holland Avenue
Box 3535
Lancaster, Pennsylvania 17604 U.S.A.

Distributed in the Rest of the World by
Technomic Publishing AG

Printed in the United States of America
10 9 8 7 6 5 4 3 2 1

Main entry under title:
Fatigue of Composite Materials

A Technomic Publishing Company book
Bibliography: p.
Includes index p. 181

Library of Congress Card No. 86-51530
ISBN No. 87762-516-6

Table of Contents

PART TWO: FATIGUE DAMAGE CHARACTERIZATION

PART THREE: FATIGUE RELIABILITY

Foreword

When the application of a material to an engineering component is contemplated, it is essential to answer not only the fundamental questions regarding the strength and stiffness of the material, but also the question of how long the material will last under the conditions anticipated. While the concept of life of engineering components is difficult to define in a general sense, the reduction of strength and the subsequent failure of materials subjected to repeated loading has been addressed as a fundamental problem since the early 1800s and is popularly known as the phenomenon of fatigue. Since a majority of engineering components are subjected to load and environmental histories which vary in time over the period of service of the component, it is not surprising that this subject has occupied the attention and the resources of a great number of scientists and engineers over the period of about 150 years.

A rational description of fatigue based on first principles has not yet been achieved. Current descriptions rely heavily on experience reflected in empirical and phenomenological descriptions of the behavior of familiar materials, aided by the subdiscipline of fracture mechanics and related formulations. In recent years, the burgeoning technology of composite materials has opened new horizons to scientists and engineers. The use of these materials is rapidly expanding, especially in high performance structures such as aircraft and other vehicles. Indeed, the applications for which composite materials are being found to be most advantageous are precisely those situations in which the degradation of strength and life by fatigue processes is most likely. Hence, it is critical that this problem be addressed for composite materials. Failure of composite structures under long-term service must be avoided; but nothing can be reliably avoided which cannot be reliably predicted. Such prediction requires understanding and quantitative representation.

This is indeed a demanding task for composite materials. By their fundamental nature, composite materials are always inhomogeneous and frequently anisotropic. The fatigue processes which reduce strength in these materials are generally very complex, involving the accumulation of multiple damage modes which may combine in a variety of ways to produce numerous failure modes. Hence, it is essential that this subject to addressed for composite materials by an interdisciplinary approach which combines a thorough and precise understanding of the physical mechanisms that play an active role in these processes and a rigorous representation of the mechanics associated with their collective action.

This interdisciplinary approach to the subject of the fatigue of composite materials is a particularly distinctive feature of the present book. The treatment begins with a description and discussion of the physical mechanisms which are responsible for the fatigue process in composite materials of various types. Conceptual formulations of the roles played by the reinforcement, the matrix, and interfacial regions are developed with emphasis on fibrous composite materials, especially those with polymer, glass, and ceramic matrix systems. The influence of these processes on the properties, especially stiffness, is discussed in some detail. The book follows with a continuum mechanics characterization of damage which incorporates distinguishing features of the damage into the material response functions of the material as a set of internal vector field variables. The constitutive equations for isothermal, small-deformation behavior of composite materials in a given damage state are then derived and the characterization is verified by deriving residual elastic properties such as stiffness in the presence of damage, and comparing with experimental data. While the issues of residual strength and life are not explicitly addressed by this approach, a solid foundation is formed on which such an enterprise could be constructed. Finally, the book addresses the subject of fatigue reliability, especially the estimation of Weibull parameters and the use of such representations for design criteria for composite structures. Strength degradation models are proposed for reliability analysis.

The book should be of considerable use to a wide range of readers. Designers and others charged with the responsibility of constructing composite structures will find the book to be a valuable summary of the nature of fatigue in composite materials and a useful indicator of problem areas and solution schemes that can be found either in its pages or in the references noted. Students of the field will find the book to be a very useful summary of fundamental information which covers an unusually wide scope, and will profit from the insights provided from Talreja's experience and expertise. In all, the book is an important contribution to an essential field of study. It is a carefully and precisely written, thoroughly annotated, and rigorously developed presentation of an important body of material having to do with the fatigue of composite materials.

Kenneth L. Reifsnider
Reynolds Metals Professor of Engineering Science and Mechanics
Virginia Polytechnic Institute and State University
Blacksburg, Virginia 24061-4899 USA

Preface

This research monograph reports the author's work related to fatigue of composite materials accomplished during the period 1979–84.

The work has been influenced by two active research environments: Department of Solid Mechanics, The Technical University of Denmark, and Metallurgy Department, Riso National Laboratory. The dual influence has been instrumental in inspiring the author to take an interdisciplinary approach combining concepts from mechanics and materials science. Further impetus to the work has come from the Materials Response Group, Virginia Polytechnic Institute and State University, and in particular from the group's chairman, Professor K. L. Reifsnider, whose long-standing collaboration and friendship has been of great benefit to the author.

I wish to thank Professor Frithiof Niordson, Head of the Department of Solid Mechanics, for his encouragement, support and influence. Dr. Techn. Niels Hansen, Head of the Metallurgy Department, Riso National Laboratory, is thanked specially for his inspiration and unfailing support during my more than five years of employment on his scientific staff.

P. Lemming, E. Dam Ravn, K. Fries Weihe and R. Zetterlund are thanked for their cooperation and expert technical assistance, and I am thankful to Bente Andersen and Lizzie Brane for their prompt and careful secretarial work.

No words would do justice to my expression of thanks to my wife and our two little ones, whose love is a source of added strength in facing the uneven path of research.

Ramesh Talreja
Lyngby, January 1985

Part 1

THE FATIGUE PROCESS

CHAPTER 1

Fatigue Damage Mechanisms: Introduction

THIS MONOGRAPH PRESENTS analyses and methods for assessing the performance of fiber-reinforced composite materials subjected to fatigue loads. The monograph is based on the author's work over the period 1979–1984.

Although the concept of fiber reinforcement is not new, and was even familiar to brickmakers thousands of years ago (see Hashin [10]), the upsurge in the interest in composite materials in the last few decades is due primarily to the high demands on materials performance placed by advanced technologies. Conventional materials are unable to meet these demands.

Improvements in the mechanical properties, e.g. stiffness, strength and fracture toughness, are being increasingly made feasible by advancements in the composite manufacturing technologies. However, before the potential of composite materials as advanced materials can be fully harnessed, a methodology for assessment of the long-term properties must be available. Development of such a methodology requires analysis of the basic mechanisms of degradation of properties under mechanical loads and incorporation of this knowledge in development of appropriate material models. The studies constituting the present monograph have been directed toward this goal.

The long-term behavior of composite materials has been the subject of active research in recent years, as testified by several symposia devoted fully or partly to this subject [11–15]. Numerous studies reported at these symposia and in the general literature have reiterated the complex nature of damage in composite materials. There has been an increasing awareness of the fact that analyses of composite damage must take an interdisciplinary approach, combining the physics of damage and the mechanics of solids. In fact, a scientific discipline for studies of such nature has recently surfaced under the name, "Damage Mechanics," and is gaining attention. The present work is intended as a contribution to this discipline.

The Fatigue Process

A number of recent reviews have presented state-of-the-art understanding of damage mechanisms in composite fatigue [16–19]. The present author's work, reported in Reference [1–2], has presented a systematic conceptual scheme for evaluation of fatigue performance of composites based on certain damage-mechanisms maps, called fatigue-life diagrams, which are principally of the same nature as the well-known fracture-mechanism maps developed by Ashby [20]. In Reference [3] the author has reviewed the basic understanding of fatigue damage in composites and discussed the significance of the fatigue-life diagrams in producing a synthesized overall picture of fatigue damage.

Figure 1, reproduced from Reference [1], shows the idealized schematics of fatigue damage mechanisms in unidirectional composites under tensile loading parallel to fibers. The resulting fatigue-life diagram is shown in Figure 2, where the maximum applied strain is plotted against the logarithm of the number of cycles to failure. The scatter band centered about the composite failure strain corresponds to fiber breakage and the resulting interfacial debonding. The sloping scatter band extending from this band to the horizontal line representing the fatigue limit corresponds to matrix cracking and interfacial shear failure. The fatigue limit is defined as the strain corresponding to the boundary between the non-propagating matrix cracks and the propagating matrix cracks at 10^6 cycles.

The scatter band corresponding to fiber breakage is taken horizontal, assuming no strength degradation due to this mechanism. Furthermore, this damage is characterized as non-progressive in the sense that a damage zone growing from an early stage of fatigue life to the final failure does not exist. The progressive damage is represented by the sloping scatter band for which the underlying mechanisms are matrix cracking and interfacial shear failure. At low strains, matrix cracking is the only operative mechanism, and the strain below which the matrix cracks remain non-progagating at 10^6 cycles is taken as the fatigue limit.

It should be emphasized here that the mapping of damage mechanisms into the fatigue-life diagram is largely a conceptual exercise. A more quantitative model-based mapping would require knowledge of the rate-controlling parameters, which is not yet fully developed.

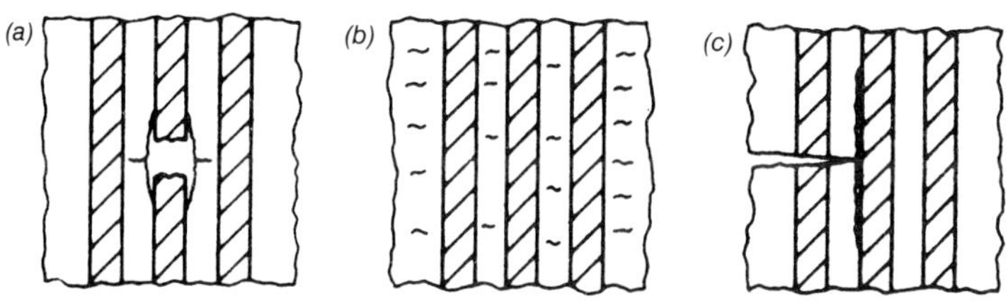

Figure 1. Fatigue damage mechanisms in unidirectional composites under loading parallel to fibers; a) fiber breakage, interfacial debonding; b) matrix cracking; c) interfacial shear failure.

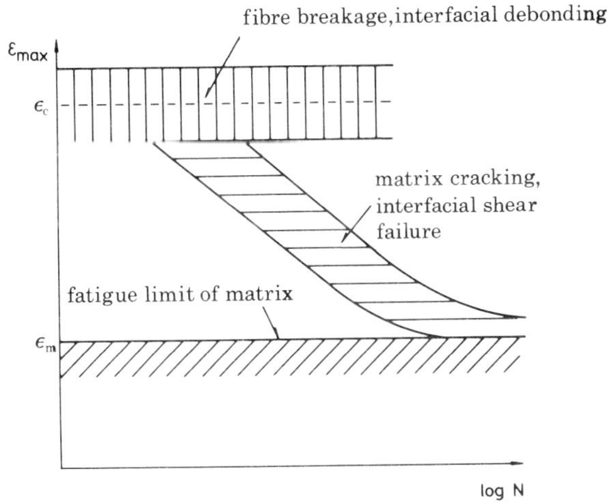

Figure 2. Fatigue-life diagram for unidirectional composites under loading parallel to fibers.

In Chapter 2 [1], the usefulness of the proposed fatigue-life diagram in evaluating the fatigue properties of unidirectional composites is demonstrated. Fiber stiffness is taken as a parameter for elucidating the relative roles of the damage mechanisms in the total fatigue behavior. In particular, it is demonstrated that the hitherto unexplained lack of fatigue damage in high modulus graphite/epoxy composites is an expected consequence of the fiber stiffness in accordance with the fatigue-life diagram.

Fatigue of unidirectional composites under loads inclined to the fiber axis is considered next in Chapter 2. Taking the mixed-mode cracking of the matrix as the operative mechanism, the evolution of the fatigue-life diagram is described. Figure 3 illustrates this evolution. Variation of the fatigue limit with the off-axis angle is illustrated using data of Hashin and Rotem [21], Figure 4 and Figure 5 show the enhancement of the fatigue limit due to the influence on the matrix cracking of the neighbouring plies in symmetric angle ply laminates. The test data used here is from Rotem and Hashin [22].

Fatigue-life diagram for cross-ply laminates is discussed next in the second chapter. Fiber breakage in longitudinal plies is assumed as the non-progressive damage, as for the unidirectional composites. Matrix cracking along fibers (transverse fiber debonding) in the transverse plies and the consequent delamination between the longitudinal and the transverse plies are taken as progressive mechanisms. Finally, laminates with combinations of 0°, 45° and 90° plies are considered and the fatigue-life diagram for this class of laminates is found to have the same upper and lower limits as the cross-ply laminate.

Fatigue ratio is proposed in terms of strains and is defined as the ratio of the fatigue limit to the static fracture strain. This ratio is a useful index for rating fatigue properties of composites. As useful guidelines for designers,

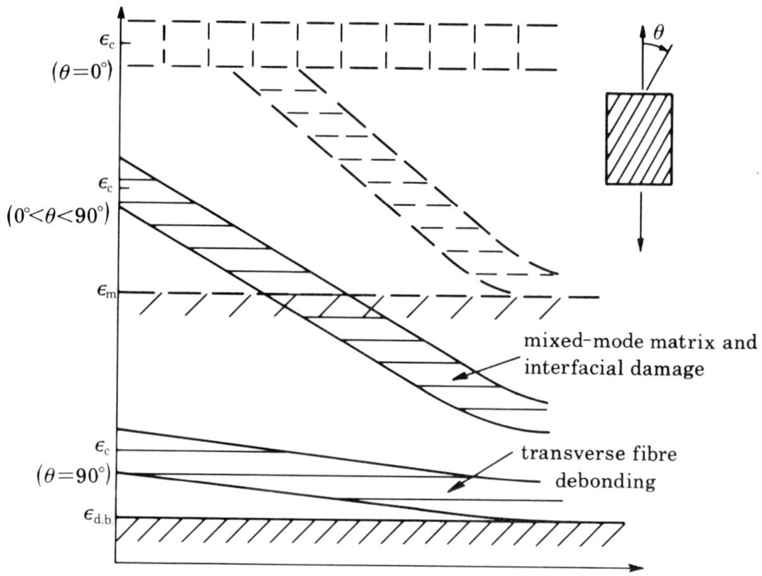

Figure 3. Fatigue-life diagram for off-axis fatigue of unidirectional composites. Dotted lines correspond to the fatigue-life diagram for on-axis fatigue.

typical values of fatigue limits for various epoxy-based composites are given. The role of matrix properties in composite fatigue resistance is also discussed.

Chapter 3 [2] generalizes the analysis presented in the second chapter to include unidirectional composites of brittle matrices such as glass and glass-ceramics. Mechanisms of damage in matrix, fibers and interface are considered separately and their relative significance is discussed in the contexts of a ductile matrix and of a brittle matrix. A fatigue-life diagram is proposed for brittle matrix composites loaded along the fiber axis.

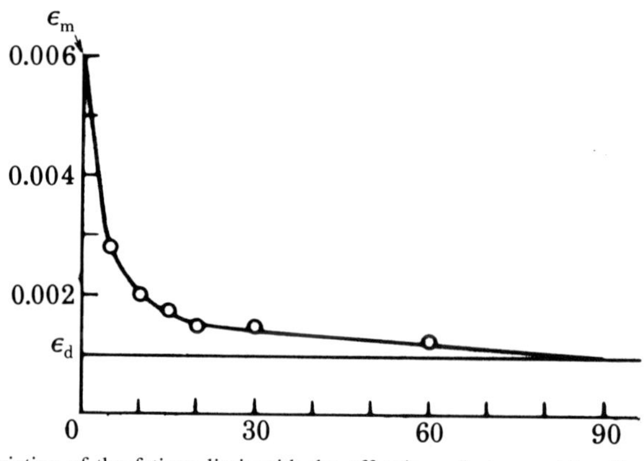

Figure 4. Variation of the fatigue limit with the off-axis angle in a unidirectional composite.

Figure 6 shows variation of the fatigue-life diagram with fiber volume fraction of brittle matrix composites. The fiber volume fraction is divided in three ranges on the basis of the characteristic values V_1 and V_2. V_1 represents the limiting value below which cracking in the matrix initiates at the failure strain of the unreinforced matrix. V_2 is the other limiting value, beyond which the matrix cracking is fully suppressed by the fibers. The figure shows typical features of the fatigue-life diagram for each range. In the first range, the strength degradation is limited by the failure strains of fibers and matrix, and is not affected by the fiber volume fraction. In the second range, the fiber volume fraction determines the strain to onset of multiple cracking in matrix, ϵ_{muc}, and thereby limits the range of strength degradation. The strength degradation is absent in the third range.

The validity of the proposed fatigue-life diagram is also sought in Chapter 3. Test data on fatigue of brittle matrix composites is very limited due to difficulties of manufacture. The available data for flexural fatigue of graphite fiber reinforced glass is utilized and it is found that the general features of the proposed fatigue-life diagram are supported by the data.

A review is presented in Chapter 4 [3] outlining the historical development of the understanding of fatigue damage mechanisms in composites. The role of fatigue-life diagrams proposed in the second and third chapters in providing a conceptual framework for interpretation of fatigue damage mechanisms is highlighted.

In concluding this section it is emphasized that the work reported in the next three chapters is an attempt to synthesize to create an overall picture of the fatigue process relating the roles of the individual damage mechanisms. One outcome of the effort is a clarification of the limits to fatigue performance of composites. Although the proposed fatigue-life diagrams do not provide life predictions, as rightly noted by Kelly in a review of composite principles [23], the diagrams give a rational basis for formulating predictive models.

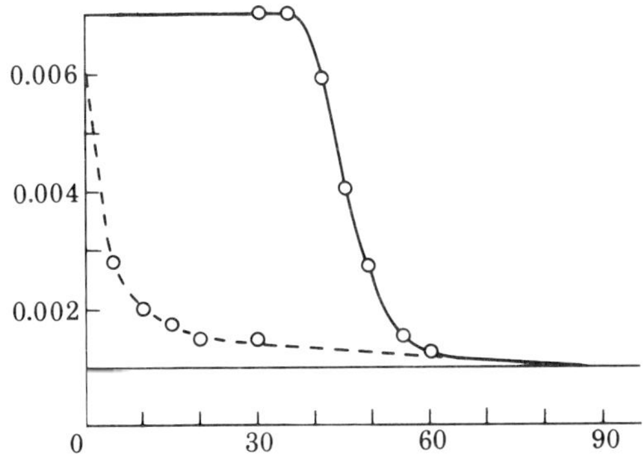

Figure 5. Variation of the fatigue limit with the fiber angle in a symmetric angle ply laminate. Dotted curve shows variation of the fatigue limit with the off-axis angle in a unidirectional composite.

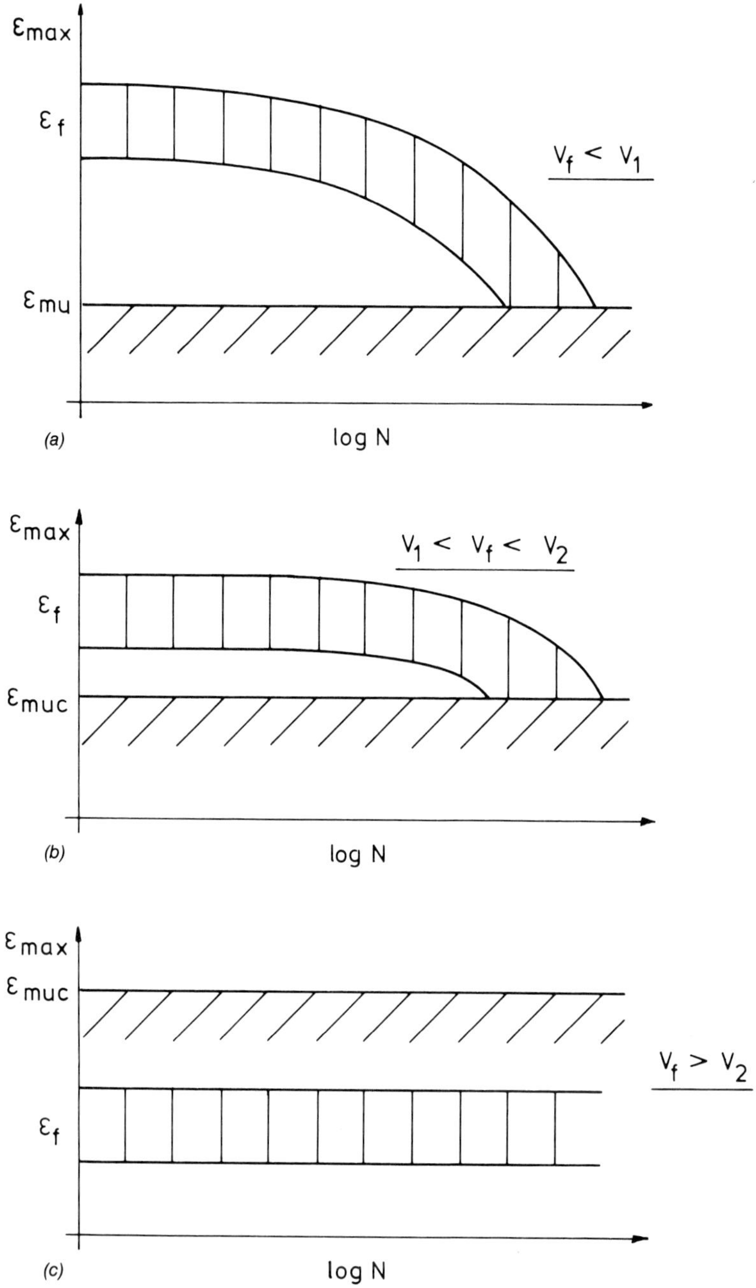

Figure 6. Fatigue-life diagram for brittle matrix composites for three ranges of fiber volume fraction.

Fatigue Damage Characterization

This section deals with characterization of the state of damage of composite materials and its relation to the overall mechanical response. Of particular interest are the relationships between stiffness properties and the interior damage in composites. The work reported in Chapters 5–8 [4–7] is connected with this subject.

Several experimental studies in the late 60s and the early 70s observed reduction of the elastic modulus in the loading direction during fatigue (see review by Salkind [16]). This spurred formulation of models, primarily of the unidirectional type, e.g. the gradual failure model by Hahn and Tsai [24], that attempted to relate the modulus degradation to the interior damage. Also, a fatigue failure criterion based on the longitudinal modulus reduction was proposed by Hahn and Kim [25].

Reifsnider [26] argued that the nature of fatigue damage in composites was anisotropic and therefore all stiffness properties would be affected. On this premise, O'Brien and Reifsnider [27] developed an experimental procedure for measurement of four independent moduli of an orthotropic composite specimen. Their procedure used a combination of uniaxial tension, rail shear and flexure tests on a specimen subjected to fatigue damage. The present author devised a simpler procedure for measurement of the moduli that relies largely on the test set-up for fatigue. This is reported in Chapter 5 [4].

The procedure developed in Chapter 5 utilizes a strain gauge rosette of $3 \times 120°$ placed symmetrically at the center of the specimen. For measurement of the moduli following a number of fatigue cycles, strains are recorded during a longitudinal load and during a load applied at 15° to the longitudinal axis. For the off-axis loading, a loading fixture is designed which requires removing the specimen from the longitudinal loading mode, tilting it by 15° and replacing it in the loading fixture. The longitudinal elastic modulus, the major Poisson's ratio and the shear modulus are given by the two tests. To obtain the minor Poisson's ratio, the specimen is to be clamped laterally and the strains recorded. It should be pointed out that a true value of the Poisson's ratio may not be given by this procedure. However, it is believed that the relative change of this property given by the measurements can be relied upon. The procedure is applied to specimens of unidirectional glass/polyester and it is found that the longitudinal Young's modulus and the major Poisson's ratio change slightly, while the shear modulus and the minor Poisson's ratio change significantly during fatigue.

Development of a workable procedure for measurement of elastic moduli has been crucial to the further progress in relating stiffness changes to the underlying damage. Reifsnider and co-workers have reported several stiffness change measurements, e.g. for boron/epoxy laminates [26] and for glass/epoxy laminates [27]. These researchers have also measured crack densities in individual plies and have correlated these with the observed stiffness loss. Sufficient evidence that stiffness change is an indicator of damage in composites has indeed been established.

Awareness of the importance of stiffness in characterization of damage in composites has led to intensified search for analytical relationships between stiffness and damage. Among the first attempts toward this was the work of Gottesman, Hashin and Brull [28], who considered a lamina in plane stress with cracks parallel to fibers. Assuming the cracks to be non-interacting and using expressions for the elastic energy associated with a single crack in an infinite orthotropic sheet, they derived expressions for the overall elastic constants. They also gave estimates for the effective elastic constants by the self-consistent scheme and provided bounds to these by a variational method. If one uses their calculations in conjunction with the laminate analysis, a procedure for estimating laminate stiffness changes due to ply cracking would be available. However, before confidence can be placed in such a procedure, an experimental verification would be necessary.

Another method for calculating stiffness reductions in unidirectional composites due to matrix cracking has been given more recently by Laws, Dvorak and Hejazi [29]. They assumed the elastic moduli of a lamina with matrix cracks to be given by the effective moduli of an infinite medium with the same fiber volume fraction and crack density as in the cracked lamina. The overall elastic moduli of the infinite medium were estimated by a self-consistent method assuming the medium to have three phases: matrix, elliptical fibers and elliptical voids, which in the limit of the aspect ratio tending to zero were taken to model the cracks. Numerical estimates of stiffness changes for different fiber volume fractions and crack densities were given. Later, the scheme was combined with the laminate analysis to produce estimates of the laminate stiffness reductions with cracking in individual plies [30].

In Chapter 6 [5] an entirely different approach for determining stiffness-damage relationships is presented. Unlike previous approaches, which may be characterized as micromechanics modelling of damage, the present approach adopts an internal state variable representation of damage. The composite damage is here looked upon as the total set of microstructural changes or rearrangements that may be divided into subsets, called damage modes, having distinct geometrical properties and growth characteristics. Examples of damage modes are matrix cracking in a given ply, interior delamination and fiber microbuckling.

In selecting the internal variables, consideration is given to the following properties of composite damage which are thought to be essential for determining the mechanical response.

1. Each damage mode consists of multiple microcracks (or microfailures). A single dominant crack, such as in the case of metals, does not exist.
2. Damage initiates and develops along preferred orientations.

A scalar variable representation of damage is not considered adequate since it would not incorporate the orientation effects. A general tensorial representation would risk rendering a phenomenological constitutive theory, in which the representation may be used, impracticable by requiring determination of a large number of phenomenological constants. A vectorial representation is thought to be a desirable choice, since it would have the capability of in corporating the two above-mentioned properties of composite damage.

A vectorial representation of damage is proposed in the form illustrated by Figure 7. A vector $\mathbf{p}^{(r)}$ is assigned to a microcrack numbered r in a family of m cracks contained by an elemental volume dv about a pt. P in a composite. The orientation of the vector is taken normal to the plane of the crack and its magnitude is taken as a function of the area and a characteristic dimension of the crack. Assuming n damage modes, i.e. n sets of oriented cracks, damage vectors $\mathbf{V}^{(\alpha)}$, $\alpha = 1, 2, \ldots n$, are defined. Each vector $\mathbf{V}^{(\alpha)}$, (oriented in the αth direction) is assigned a magnitude which is an average of the magnitudes of all vectors $\mathbf{p}^{(r)}$ oriented in the αth direction.

The vectors $\mathbf{V}^{(\alpha)}$ are taken as a set of internal state variables in the phenomenological constitutive equations formulated by Coleman and Gurtin [31]. Restricting consideration to isothermal, small strain behavior, the material response is characterized by the Helmholz free energy, which is here a function of the strain tensor and the set of damage vectors.

Assuming the Helmholz free energy as a polynomial function of its variables, restrictions to its form placed by the material symmetry properties are sought. Two types of symmetry of interest to composite materials, namely, the transversely isotropic symmetry and the orthotropic symmetry are considered. The restricted forms of the Helmholz free energy function are developed using the appropriate invariants given by Rivlin and coworkers.

The stress-strain equations derived from the Helmholz free energy show that the stiffness matrix consists of two parts: one part related to the undamaged response and the other part due to damage. The equations show

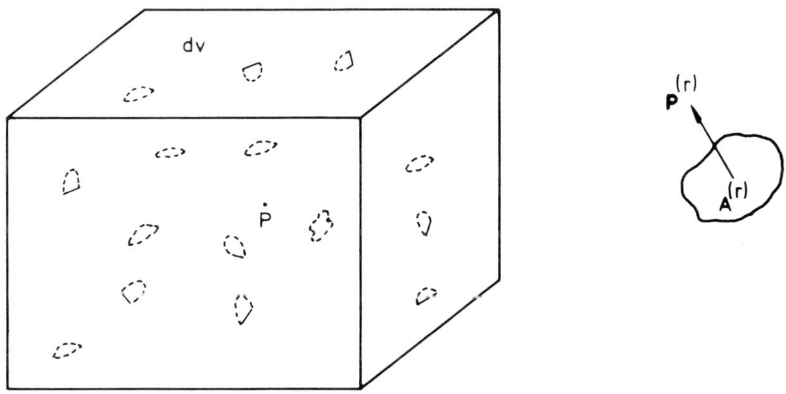

Figure 7. A vectorial representation of damage.

further that admittance of any damage removes the initial symmetry of a composite. Specific cases of damage are considered for illustrating the changes of stiffness and symmetry brought about by damage.

For low concentration of damage, equations relating the residual moduli to the initial moduli and the damage vector magnitudes are given. These equations show that, even for a single damage mode, all elastic moduli change. The generality of the equations is demonstrated by comparison with the effective elastic moduli derived by Gottesman, Hashin and Brull [28] for a cracked lamina. It is shown that the Gottesman-Hashin-Brull expressions are a special case of the residual moduli equations derived here.

The paper discusses the limitations of the work related to description of damage and to characterization of the evolution of damage. In particular, the relevance of the stiffness-damage relationships to characterizing fatigue damage is discussed.

In Chapter 7 [6] attention is focussed on a special case of great practical value. Thin orthotropic laminates having matrix cracks in individual plies are considered. A damage vector associated with cracking in a given ply is assigned a magnitude given by

$$\mathbf{V} \cdot \mathbf{V} = D^2 = \eta_c \cdot \bar{\ell}_c \cdot \bar{w}_c \cdot f_c$$

where η_c is the crack number density, i.e. the number of cracks per unit surface area of laminate, $\bar{\ell}_c$ and $\bar{w}_c$ are the average length and the average width of cracks, respectively, and f_c is a factor depending on crack shape and on the constraint to crack surface displacements from the surrounding material.

Restricting to inplane deformation and inplane damage vectors, the elastic response is characterized by the strain energy function, which is expanded as a polynomial of the invariants associated with the orthotropic symmetry of the laminate. For small magnitudes of damage vectors, stress-strain equations are derived which incorporate the effect of cracks on the stiffness co-efficients. Expressions for the residual moduli due to one set of cracks lying parallel to a plane of symmetry take the following form

$$E_1 = E_1^\circ + 2D^2\left[k_3 + k_7\left(\nu_{12}^\circ\right)^2 - k_{13}\nu_{12}^\circ\right]$$

$$E_2 = E_2^\circ + 2D^2\left[k_7 + k_3\left(\nu_{21}^\circ\right)^2 - k_{13}\nu_{21}^\circ\right]$$

$$\nu_{12} = \nu_{12}^\circ + D^2\left[\frac{1 - \nu_{12}^\circ\nu_{21}^\circ}{E_2^\circ}\right]\left(k_{13} - 2k_7\nu_{12}^\circ\right)$$

and

$$G_{12} = G_{12}^{o} + 2D^2k_{11}$$

where the moduli with superscript zero are the initial moduli and k_i are material constants.

The procedure for determining k_i consists of solving for the preceding equations using values of the initial moduli and (measured) values of the residual moduli for a given damage vector magnitude D. The procedure is demonstrated in Chapter 7 by use of experimental data reported by Highsmith and Reifsnider [27]. In calculating D^2 given by the formula above, the average crack length is taken as the distance spanned by the cracks from one edge of the specimen to another, the average width is taken as the thickness of plies containing cracks and the factor f_c is taken simply as the ratio of the average crack width to the laminate thickness.

The constants k_i are determined for one laminate configuration of glass/epoxy and thereafter used for predicting moduli changes for other laminate configurations of glass/epoxy Table 1 summarizes the comparison of the predicted and observed moduli reductions. The shear modulus is not included because of the uncertainty of measurement of this property in the experimental procedure used [27]. However, since the shear modulus is uncoupled from the remaining moduli for orthotropic damage (see equations above), the predictions in Table 1 are a sufficient check of the accuracy of the procedure for determining moduli reductions.

In Chapter 8 [7] transverse cracking in plies and its effect on laminate stiffness is studied further. In particular, the effect of constraint to transverse cracking on the consequent stress-strain behavior is considered. The constraint under question depends on two quantities: the thickness of the cracking plies, which sets the limit to the size of cracks, and the stiffness of the plies neighbouring to the cracking plies. The thickness effect is illustrated by considering the variation of the strain to onset of transverse cracking in a cross-

Table 1. Predicted and observed moduli reductions for glass/epoxy laminates. Laminate 1: $(0,90_3)_s$, Laminate 2: $(90_3,0)_s$, Laminate 3: $(0,90)_s$, Laminate 4: $(0,\pm45)_s$.

	ΔE_1, %		$\Delta\nu_{12}$, %		ΔE_2, %	
Laminate	Observed	Predicted	Observed	Predicted	Observed	Predicted
2	38.8	38.1	71.0	67.4	0.0	0.5
3	17.1	17.2	49.8	41.4	7.3	0.4
4	9.4	8.3	8.1	7.5	5.3	0.1

$k_3 = -6.713$ GPa, $k_7 = -0.762$ GPa and $k_{13} = -4.467$ GPa. Data for Laminate 1: $\Delta E_1 = 42.0\%$, $\Delta\nu_{12} = 74.3\%$ and $\Delta E_2 = 0.6\%$.

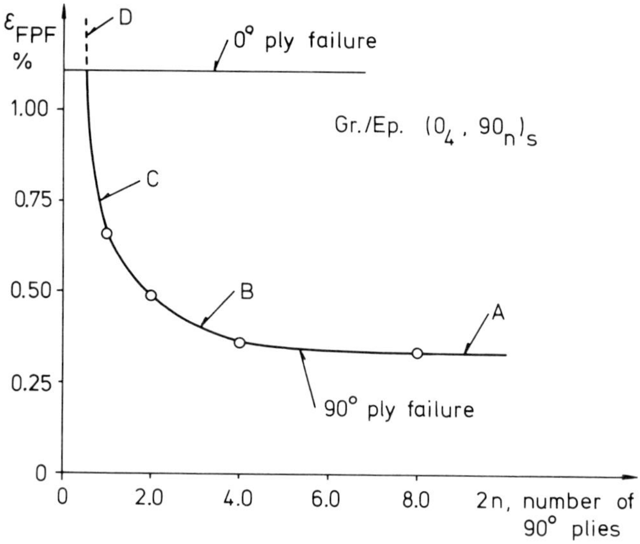

Figure 8. Variation of the strain to onset of transverse cracking, ϵ_{FPF}, with the thickness of the transverse plies.

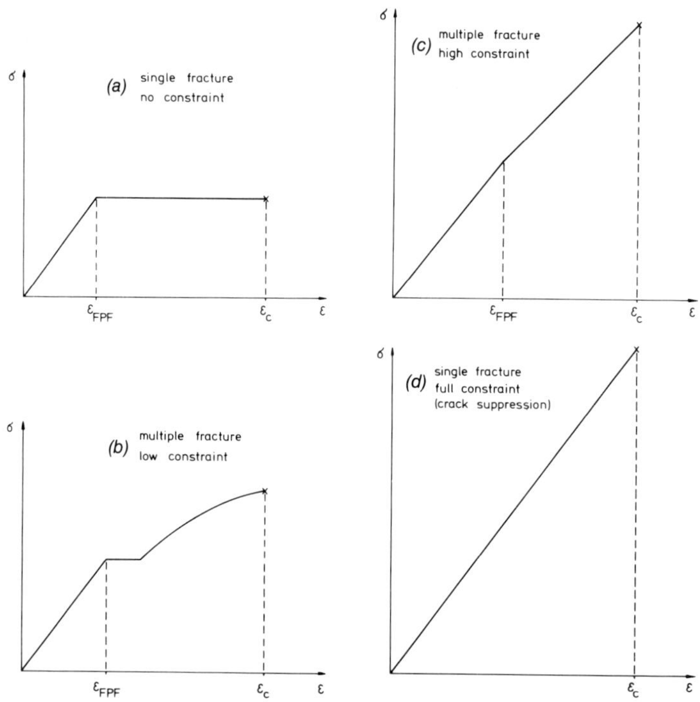

Figure 9. Schematic stress-strain diagrams for varying degrees of constraint to transverse cracking.

ply laminate in which the longitudinal plies are kept fixed while the number of the transverse plies is varied. Figure 8 shows this variation. Four regions are qualitatively defined on this curve to characterize the degree of constraint as no constraint (A), low constraint (B), high constraint (C) and full constraint (D). The stress-strain behavior corresponding to these regions is illustrated schematically in Figure 9.

Experimental data concerning transverse crack densities, stress-strain behavior and moduli reductions are taken from the literature to illustrate the constraint effects and to compare moduli predictions with observed values. For laminates of glass/epoxy moduli, reductions are predicted from onset of transverse cracking to its saturation both under static loading and under fatigue loading. Two examples are shown in Figure 10 and Figure 11. Laminates of graphite/epoxy of different cracking ply thicknesses and different constraining ply orientations are also considered. Two examples of moduli predictions are shown in Figure 12–13. Figure 12 shows variation of the longitudinal Poisson's ratio of a $(0,90)_{4s}$ laminate with the applied static stress. Figure 13 shows variation of the longitudinal Young's modulus of a $(45,90,-45,90,45,90,-45,\overline{90})_s$ laminate.

Along with prediction of moduli reductions by the present method, an assessment is made of an engineering scheme for moduli prediction called the ply discount method. In the commonly used version of this scheme, the transverse and shear moduli of a cracked ply are reduced to zero and the laminate analysis is used to recalculate the laminate stiffnesses. Estimates of the minimum values of the reduced moduli are given by this scheme. These are usually taken to correspond to the state of crack saturation in a laminate. Highsmith and Reifsnider [27] have used a modified version of the scheme by setting only the transverse modulus of a cracked ply to zero. This makes a difference only when a cracked ply has an orientation other than 90° with respect to the longitudinal axis.

The results of Chapter 8 show that the ply discount method is not generally reliable. It appears to overestimate moduli reductions for high constraint cracking and underestimate these for low constraint cracking. This trend needs however to be verified by more results.

The work described in Chapters 5–8 provides collectively a methodology for characterizing damage and the consequent response of composite materials under mechanical loads, in particular under fatigue. The approach taken is based on well-founded solid mechanics principles and the procedure for quantitative prediction of moduli reductions is found to be practical and of acceptable accuracy.

Fatigue Reliability

Fatigue damage in composite materials is a process of stochastic nature. Consequently, probabilistic concepts and methods are needed for an assess-

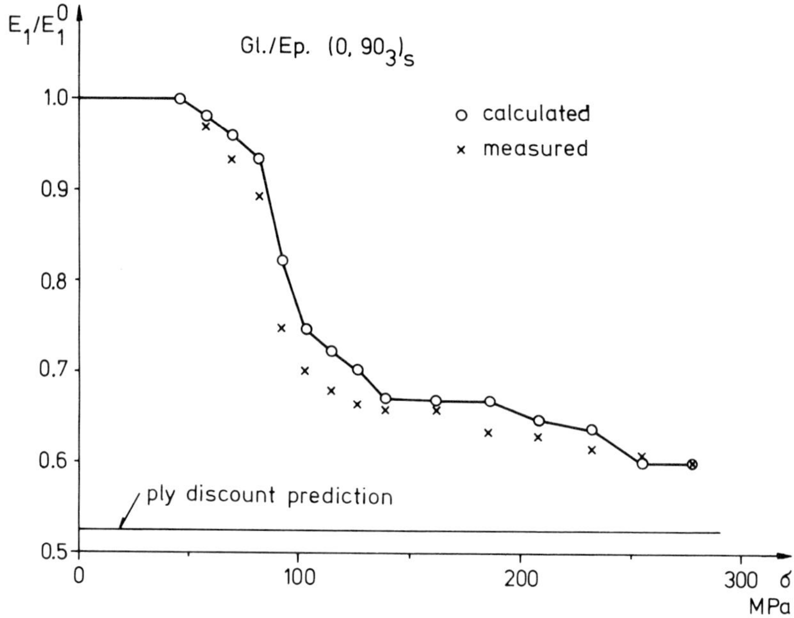

Figure 10. Variation of the longitudinal Young's modulus of $(0,90_3)_s$ glass/epoxy laminate with the applied static stress.

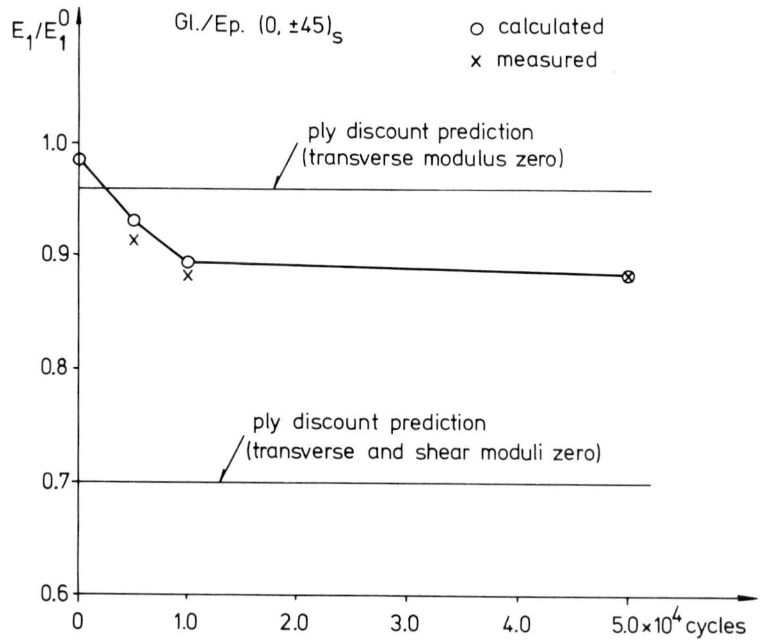

Figure 11. Variation of the longitudinal Young's modulus of $(0,\pm 45)_s$ glass/epoxy laminate with the number of cycles.

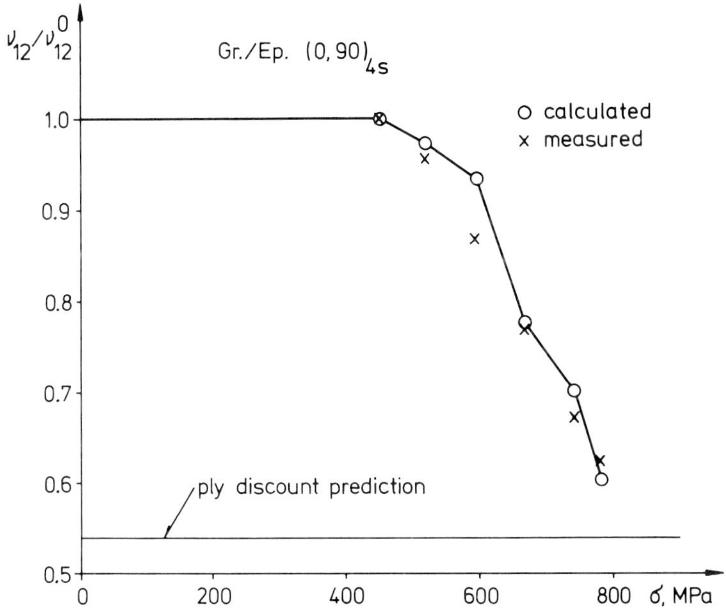

Figure 12. Variation of Poisson's ratio of a $(0{,}90)_{4s}$ graphite/epoxy laminate with the applied tensile stress.

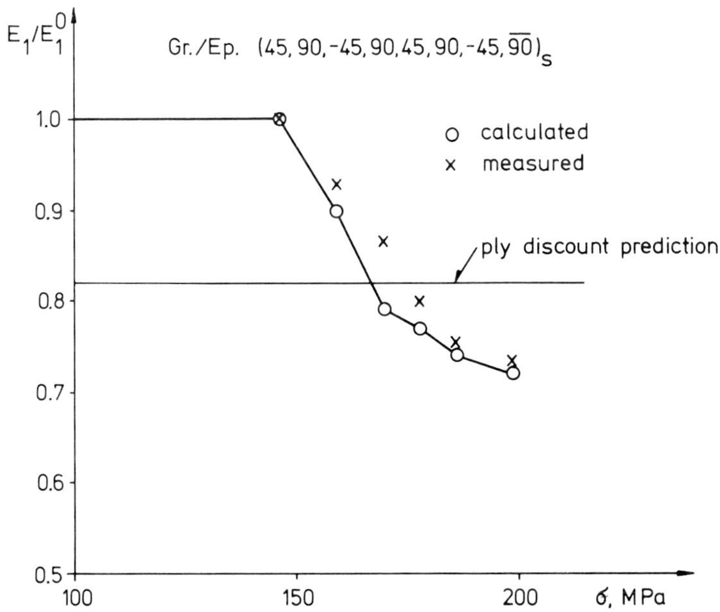

Figure 13. Variation of the longitudinal Young's modulus of a $(45{,}90,-45{,}90{,}45{,}90,-45,\overline{90})_s$ graphite/epoxy laminate with the applied tensile stress.

ment of the safety and reliability of composite structures under fatigue. Such concepts and methods have been developed in the context of metallic structures. For composite structures, modifications to these must be made to account for the characteristic features of composite damage. Several studies have recently dealt with this subject [32–38]. The present author's contributions are aimed at two aspects of fatigue reliability of composites. They form Chapter 9 [8], the estimation of Weibull distribution parameters for composite material strength and fatigue life data, and Chapter 10 [9] strength degradation model for composite fatigue damage.

Most studies of structural safety and reliability make use of the Weibull distribution for describing the statistical variability of strength and of fatigue life. Application of analyses developed in the studies often requires estimation of the distribution parameters from test data. Composite material test data are often found to display heterogeneity due to the underlying multi-modal damage process. This necessitates an estimation procedure that is capable of detecting heterogeneity in test data, of separating components and of estimating parameters of individual components. In Chapter 9 an estimation procedure of such capability is described for a three-parameter Weibull distribution. The efficiency of the procedure is demonstrated on random samples, and actual test data is used to illustrate its applicability.

For reasons of simplicity and expediency a two-parameter Weibull distribution is sometimes assumed implying thereby the location parameter of the distribution to be zero. The consequences of this assumption are also examined in Chapter 9. It is shown that for a sample belonging to a three-parameter Weibull distribution, the assumption of two-parameter Weibull distribution leads to estimates of the two parameters being higher than their true values. Furthermore, the errors made in such estimation increase linearly with increasing value of the true location parameter.

The classical methods of parameter estimation applicable to a two-parameter Weibull distribution, namely, the moment estimation method and the maximum likelihood method are reviewed. These methods are used as references for comparison with the proposed estimation procedure. The proposed estimation method, referred to as the standardized variable estimation method, is demonstrated to give accurate and efficient estimates of all three Weibull parameters. The estimates of the scale and shape parameters given by this method are found to be more accurate and more efficient than those given by the classical methods for high shape parameters.

Figure 14 shows an example illustrating application of the standardized variable estimation method. Compression strength values, x_i, of a composite laminate are plotted against median values of the standardized variable, Mz_i, for a selected shape parameter. The plot shows that the data cluster about two straight lines, indicating two separate components in the distribution. The best-fit straight lines provide estimates of the parameters for the two components.

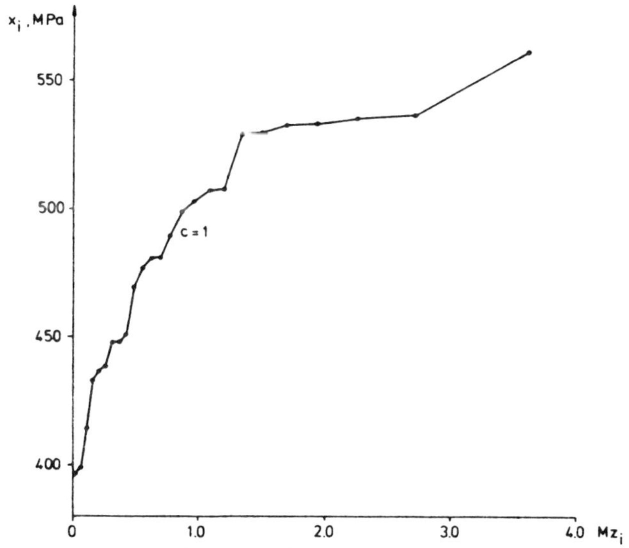

Figure 14. Compression strength data, x_i, for a composite laminate plotted against median of the standardized variable, Mz_i.

Chapter 10 deals with strength degradation of composite materials under fatigue. A commonly adopted approach to fatigue reliability assumes that the static strength of a material reduces under fatigue and failure occurs when the residual strength equals the maximum applied stress. This approach has been successfully used for predicting fatigue reliability of metals [39–41]. The strength degradation in these studies has been assumed to result from fatigue crack growth [39] and in [40–41] from microcracking as well as fatigue crack growth.

In fatigue reliability studies of composites, e.g. in [34], it has been assumed that the strength degradation results from a fatigue damage process, which is not specified in any detail. The rate of strength degradation has been taken to be inversely proportional to some power of the residual strength and, on this basis, a relationship between the initial strength and the residual strength has been derived. It is felt that a model of this nature may not be realistic, although it may seem to work for the purpose. In Chapter 10, therefore, an attempt is made to derive a strength degradation model from considerations of the fatigue damage development process in composites.

The present understanding of the fatigue development process in composite laminates, illustrated in Figure 15 taken from Reifsnider et al. [42], forms the basis for the strength degradation model proposed in Chapter 10. Two characteristic properties of major significance emerge from this understanding.

1. Damage evolves in two dominant stages. The first stage consists of homogeneous, non-interactive cracking restricted to individual plies. The second

stage is characterized by the tendency of localization of damage in zones of increasing crack interaction.

2. Transition from the first stage to the second stage occurs at the so-called Characteristic Damage State, CDS, which consists of a well-defined crack pattern characterizing saturation of the non-interactive cracking.

Strength degradation is assumed also to take place in two stages reflecting the development of the underlying damage process. In the first stage, a general weakening of the material is assumed and is taken to depend on a damage parameter representing the stage of damage. A power law is assumed for the rate of increase of the damage parameter taking it as a function of an effective stress. A relationship between the residual strength and the initial strength is then derived. Determination of the probability of failure based on this relationship is discussed.

In the second stage, strength degradation is assumed to result from the localized zones of damage which are conceptually replaced by a single crack capable of releasing the same amount of elastic energy as that released collectively by the various crack growth mechanisms in all the localized zones. Residual strength is related to a characteristic dimension of the "equivalent" crack through a fracture mechanics type relationship. The evolution of damage is expressed by the rate of growth of the crack dimension, which is assumed to depend on the current state of damage given by the current crack dimension. Assuming a power law for the crack growth rate, a relationship

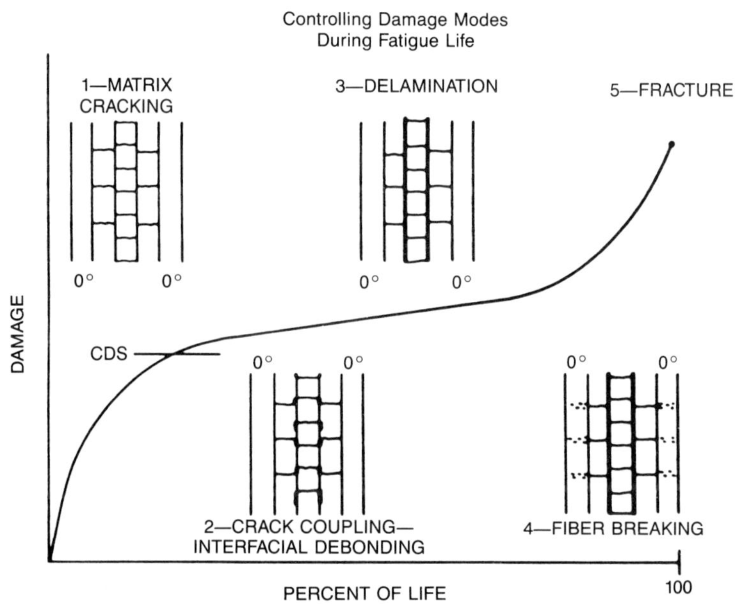

Figure 15. Development of fatigue damage in composite laminates (from [42]).

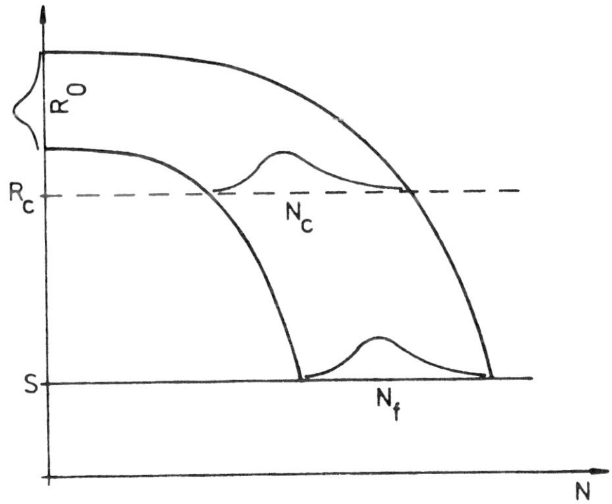

Figure 16. Two-stage strength degradation model for fatigue reliability of composites.

between residual strength and the applied maximum stress is derived. This relationship forms the basis for determining the probability distribution of the residual strength and the probability distribution of the number of cycles to attainment of the characteristic damage state.

The complete two-stage strength degradation model for fatigue reliability analysis of composites is illustrated in Figure 16 where R_o is the initial strength, R_c and N_c are the residual strength and the number of cycles corresponding to CDS, S is the maximum applied stress and N_f is the number of cycles to failure.

Conclusions

In designing structures subjected to fatigue loads, an engineer is faced with several tasks. He must select materials with desired fatigue properties, he must evaluate the mechanical response of structures damaged in fatigue and he must assure safe and reliable performance of structures for the desired fatigue life. The studies summarized above have developed methods and analyses aimed at meeting these tasks.

The studies of damage mechanisms have produced a conceptual framework for interpretation of the fatigue behavior of composites. This framework, in the form of fatigue-life diagrams, serves as a useful means of evaluating fatigue properties of candidate composite materials. Data for several polymeric matrix composites and a glass matrix composite have demonstrated the applicability of the diagrams. Also, useful design strains for epoxy matrix composites have emerged from the diagrams.

The stiffness-damage relationships developed provide an important link between the micromechanisms and the overall mechanical response of composites. The relationships are fairly general, but at the same time are amenable to special cases of damage due to their phenomenological nature. A variety of examples considered have demonstrated practicability and accuracy of the relationships in predicting stiffness changes. In addition to being instrumental in evaluating static and dynamic response, the relationships provide a means of monitoring damage in composites.

For assessment of the safety and reliability of composite structures subjected to fatigue, the assumption of strength degradation forms a basis for most studies. A probability distribution of wide applicability in this context is the Weibull distribution. The problem of estimating parameters of this distribution has been considered and useful results have been obtained. In particular, a method of good efficiency and accuracy for parameter estimation has been proposed. The applicability of the method in analyzing composite material strength and fatigue life data has been demonstrated. Finally, unlike previous studies, a strength degradation model developed from considerations of the physical mechanisms of fatigue damage development in composites has been presented.

References

1. Talreja, R., "Fatigue of Composite Materials: Damage Mechanisms and Fatigue-Life Diagrams," *Proc. Roy. Soc.*, London, A378, pp. 461–475, Reprinted below: Chapter 2 (1981).
2. Talreja, R., "Damage Models for Fatigue of Composite Materials," *Fatigue and Creep of Composite Materials*, H. Lilholt and R. Talreja, eds., pp. 137–153, Ris∅ National Laboratory, Roskilde, Denmark, Reprinted below: Chapter 3 (1982).
3. Talreja, R., "A Conceptual Framework for the Interpretation of Fatigue Damage Mechanisms in Composite Materials," *Journal of Composites Technology and Research*, 7, No. 1., pp. 25–29, Reprinted below: Chapter 4 (1985).
4. Talreja, R., "Stiffness Based Fatigue Damage Characterization of Fibrous Composites," *Advances in Composite Materials*, A. R. Bunsell, et al., eds., Vol. 2, pp. 1732–1739, Pergamon Press, Oxford, Reprinted below: Chapter 5 (1980).
5. Talreja, R., "A Continuum Mechanics Characterization of Damage in Composite Materials," to appear in *Proc. Roy. Soc.*, London, Reprinted below: Chapter 6.
6. Talreja, R., "Residual Stiffness Properties of Cracked Composite Laminates," *Advances in Fracture Research*, S. R. Valluri, et al., eds., Vol. 4, pp. 3013–3020, Pergamon Press, Oxford, Reprinted below: Chapter 7 (1984).
7. Talreja, R., "Transverse Cracking and Stiffness Reduction in Composite Laminates," to appear in *J. Comp. Mats.*, Reprinted below: Chapter 8.
8. Talreja, R., "Estimation of Weibull Parameters for Composite Material Strength and Fatigue Life Data," *Fatigue of Fibrous Composite Materials, ASTM STP 723*, American Society for Testing and Materials, pp. 291–311, Reprinted below: Chapter 9 (1981).
9. Talreja, R., "On Design Criteria for Composite Structures under Static and Fatigue Loads," *Proceedings of ICOSSAR '85*, 4th International Conference on Safety and Reliability, Kobe, Japan, pp. 425–434, Reprinted below: Chapter 10 (May 1985).

10. Hashin, Z., "Opening Remarks," *Mechanics of Composite Materials, Recent Advances*, Z. Hashin and C. T. Herakovich, eds., Pergamon Press, New York (1982).
11. *Fatigue of Fibrous Composite Materials*, ASTM STP 723, American Society for Testing and Materials (1981).
12. *Damage in Composite Materials, ASTM STP 775*, American Society for Testing and Materials (1982).
13. *Long-term Behavior of Composites, ASTM STP 813*, American Society for Testing and Materials (1983).
14. *Fatigue and Creep of Composite Materials*, H. Lilholt and R. Talreja, eds., Ris∅ National Laboratory, Roskilde, Denmark (1982).
15. *Mechanics of Composite Materials, Recent Advances*, Z. Hashin and C. T. Herakovich, eds., Pergamon Press, New York (1982).
16. Salkind, M. J., "Fatigue of Composites," *Composite Materials: Testing and Design* (Second Conference), ASTM STP 497, American Society for Testing and Materials, pp. 143–169 (1972).
17. Owen, M. J., "Fatigue Damage in Glass-Fiber-Reinforced Plastics," and "Fatigue of Carbon-Fiber-Reinforced Plastics," Chapter 7 and Chapter 8, resp., *Composite Materials 5, Fracture and Fatigue*, L. J. Boutman, ed., Academic Press, New York (1974).
18. Stinchcomb, W. W. and Reifsnider, K. L., "Fatigue Damage Mechanisms in Composite Materials," *Fatigue Mechanisms*, ASTM STP 675, American Society for Testing and Materials, pp. 762–787 (1975).
19. Hahn, H. T., "Fatigue Behavior and Life Prediction of Composite Laminates," *Composite Materials: Testing and Design*, Fifth Conference, ASTM STP 674, American Society for Testing and Materials, pp. 383–417 (1979).
20. Ashby, M. F., "Progress in the Development of Fracture-Mechanism Maps," *Fracture 1977, 1*, ICF4, University of Waterloo Press, Waterloo, Canada, pp. 1–4 (1977).
21. Hashin, Z. and Rotem, A., "A Fatigue Failure Criterion for Fiber Reinforced Materials," *Journal of Composite Materials*, pp. 448–464 (1973).
22. Rotem, A. and Hashin, Z., "Fatigue Failure of Angle Ply Laminates," *AIAA Journal*, 14, No. 7, pp. 868–872 (1976).
23. Kelly, A., "Composite Principles," *Fatigue and Creep of Composite Materials*, H. Lilholt and R. Talreja, eds., Ris∅ National Laboratory, Roskilde, Denmark, pp. 51–62 (1982).
24. Hahn, H. T. and Tsai, S. W., "On the Behavior of Composite Laminates After Initial Failures," *Journal of Composite Materials*, 8, pp. 288–305 (1974).
25. Hahn, H. T. and Kim, R. Y., "Fatigue Behavior of Composite Laminates," *Journal of Composite Materials*, 10, pp. 156–180 (1976).
26. O'Brien, T. K. and Reifsnider, K. L., "Fatigue Damage Evaluation Through Stiffness Measurements in Boron-Epoxy Laminates," *Journal of Composite Materials*, 15, pp. 55–70 (1981).
27. Highsmith, A. L. and Reifsnider, K. L., "Stiffness Reduction Mechanisms in Composite Laminates," *Damage in Composite Materials*, ASTM STP 775, American Society for Testing and Materials, pp. 103–117 (1982).
28. Gottesman, T., Hashin, Z. and Brull, M. A., "Effective Elastic Moduli of Cracked Fiber Composites," *Advances in Composite Materials*, Vol. 1, A. R. Bunsell et al., eds., pp. 749–758, Pergamon Press, Oxford (1980).
29. Laws, N., Dvorak, G. J. and Hejazi, M., "Stiffness Changes in Unidirectional Composites Caused by Crack Systems," *Mechanics of Materials*, 2, pp. 123–137 (1983).
30. Laws, N. and Dvorak, G. J., "The Loss of Stiffness of Cracked Laminates," Paper for

IUTAM/ICF/ICM Symposium on Fundamentals of Deformation and Fracture, University of Sheffield (April 2–5, 1984).

31. Coleman, B. D. and Gurtin, M. E., "Thermodynamics with Internal State Variables," *Journal of Chemical Physics*, 47, No. 2, pp. 597–612 (1967).

32. Hahn, H. T. and Kim, R. Y., "Proof Testing of Composite Materials," *Journal of Composite Materials*, 9, pp. 297–311 (1975).

33. Yang, J. N., "Reliability Prediction for Composites Under Periodic Proof Tests in Service," *Composite Materials: Testing and Design* (Fourth Conference), ASTM STP 617, pp. 272–295 (1977).

34. Yang, J. N. and Liu, M. D., "Residual Strength Degradation Model and Theory of Periodic Proof Tests for Graphite/Epoxy Laminates," *Journal of Composite Materials*, 11, pp. 176–203 (1977).

35. Yang, J. N., "Fatigue and Residual Strength Degradation for Graphite/Epoxy Composites Under Tension-Compression Cyclic Loading," *Journal of Composite Materials*, 12, pp. 19–38 (1978).

36. Yang, J. N. and Jones, D. L., "Statistical Fatigue of Graphite/Epoxy Angle-Ply Laminates in Shear," *Journal of Composite Materials*, 12, pp. 371–389 (1978).

37. Yang, J. N. and Jones, D. L., "Effect of Load Sequence on the Statistical Fatigue of Composites," *AIAA Journal*, 18, No. 12, pp. 1525–1531 (1980).

38. Hahn, H. T., "Proof-Load Determination for Pressure Vessels Wound with Aramid Fiber," *Test Methods and Design Allowables for Fibrous Composites*, ASTM STP 734, pp. 361–375 (1981).

39. Yang, J. N. and Trapp, W. J., "Reliability Analysis of Aircraft Structures under Random Loading and Periodic Inspection," *AIAA Journal*, 12, No. 12, pp. 1623–1630 (1974).

40. Talreja, R. and Weibull, W., "Probability of Fatigue Failure Based on Residual Strength," *Fracture 1977, Proc. ICF4*, Vol. 2, D. M. R. Taplin, ed., Univ. of Waterloo Press, Waterloo, Canada, pp. 1125–1131 (1977).

41. Talreja, R., "Fatigue Reliability under Multiple Amplitude Loads," *Engineering Fracture Mechanics*, 11, pp. 839–849 (1979).

42. Reifsnider, K. L., Henneke, E. G., Stinchcomb, W. W. and Duke, J. C., "Damage Mechanics and NDE of Composite Laminates," *Mechanics of Composite Materials, Recent Advances*, Z. Hashin and C. T. Herakovich, eds., Pergamon Press, New York, pp. 399–420 (1983).

CHAPTER 2

Fatigue of Composite Materials: Damage Mechanisms and Fatigue-Life Diagrams

Abstract: The basic fatigue damage mechanisms in composite laminates are reviewed. Based on these mechanisms a pattern in the fatigue-life diagrams is proposed. Several experimental data are shown to agree with this basic pattern. Fatigue ratio is defined in terms of strains, and fatigue limit is shown to exist for unidirectional, cross-plied and angle-plied laminates.

The limitations to the fatigue performance of composite laminates are pointed out and suggestions for improving the fatigue resistance are made.

COMPOSITE MATERIALS ARE increasingly believed to be the materials of the future with potential for application in high performance structures. One of the reasons for believing this is the indication that composite materials are insensitive to fatigue even at stresses close to the static fracture strength. Although this is true for advanced fibre composites subjected to tensile stresses in the fibre direction, for other stresses and fibre orientations fatigue of composites may be even more complex than metallic fatigue. Therefore, if composites are to perform better than metals, it is essential to incorporate existing knowledge of composite damage mechanisms in designing fatigue-resistant structures.

In this paper, an attempt is made to classify the composite damage mechanisms as progressive and non-progressive mechanisms, and, for each category, limiting conditions are proposed. A rational definition of fatigue ratio is suggested and fatigue limit is shown to exist in terms of strain.

The analysis presented here is limited to unnotched continuous-fibre laminates with polymeric matrix materials. The fibre orientations considered are the unidirectional, the angle-plied, the cross-plied and some combinations of these.

Reprinted by permission of The Royal Society of London, Proc. R. Soc. Lond. A378, 461–475 (1981).

Damage Mechanisms in Unidirectional Composites

Fatigue damage mechanisms in unidirectional composites depend on the loading mode, for example tensile or compressive, and on whether the loading is parallel to or inclined to the fibre direction. We shall here consider only tensile loads and discuss the damage mechanisms under parallel loading and under inclined loading separately.

FATIGUE UNDER LOADING PARALLEL TO FIBRES (ON-AXIS FATIGUE)

The damage mechanisms in tensile fatigue may be divided into three types (see Figure 1). Fibre breakage (Figure 1a) occurs at stresses exceeding the strength of the weakest fibre in the composite. An isolated fibre break causes shear-stress concentration at the interface close to the tip of the broken fibre. The interface may then fail, leading to debonding of the fibre from the surrounding matrix. The debond length depends on the shear strength of the interface and is usually small, of the order of a few fibre diameters. The debonded area acts as a stress concentration site for the longitudinal tensile stress. The magnified tensile stress may exceed the fracture stress of the matrix leading to a transverse crack in the matrix.

The fatigue mechanisms in polymeric materials are similar to those in metallic materials in the sense that they involve crack initiation and crack propagation normal to the tensile stress (Dew-Hughes & Way 1973). Under cyclic loading of a composite the constrained matrix is subjected to strain-controlled fatigue and if the applied cyclic strain exceeds its fatigue (strain) limit, the matrix cracks (Figure 1b). A matrix crack would stop at the interface at low strains, while, at high strains, the stress at the crack tip might exceed the fracture stress of fibres, leading to fibre failure. The matrix crack may now propagate under fatigue as a macrocrack in an opening mode until it hits an interface, where the shear stresses may cause its propagation in the sliding mode leading to a progressive failure of the interface (Figure 1c).

The three types of damage mechanisms may occur simultaneously. However, observations by Owen (1974) and Dharan (1975) indicate that the predominant mechanism leading to failure would be effective in a limited range

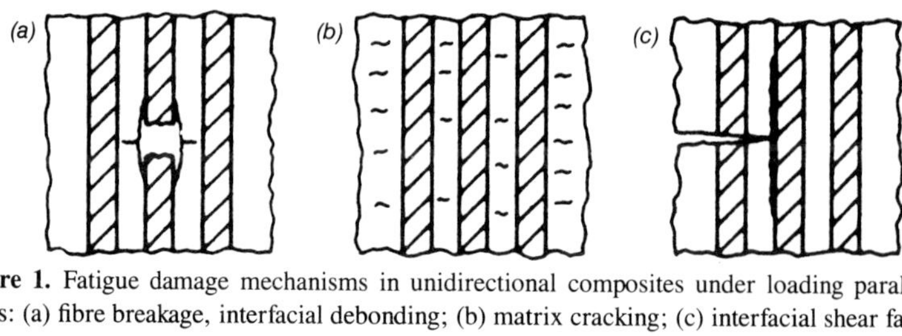

Figure 1. Fatigue damage mechanisms in unidirectional composites under loading parallel to fibres: (a) fibre breakage, interfacial debonding; (b) matrix cracking; (c) interfacial shear failure.

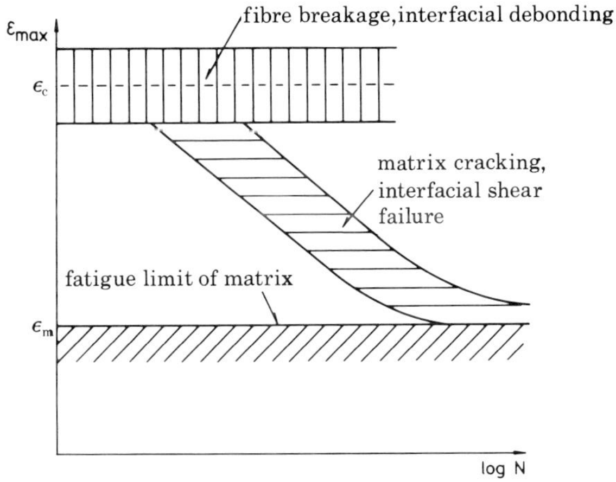

Figure 2. Fatigue-life diagram for unidirectional composites under loading parallel to fibres.

of applied maximum strain. On this assumption, it would be expected that the fatigue-life curve would consist of different components, each corresponding to the underlying damage mechanism. Based on the damage mechanisms described, a pattern for the fatigue-life diagram is suggested below.

FATIGUE-LIFE DIAGRAM

In Figure 2 a schematic fatigue-life diagram is constructed for tensile fatigue of unidirectional composites. The coordinates chosen are the maximum strain and the logarithm of the number of cycles to failure. Strain instead of stress is chosen as an independent variable, as both fibres and matrix would be subjected to the same strain while stresses in the two phases would differ depending on the volume fraction and the elastic moduli of the two phases. As shown in the figure, the horizontal band centred about the composite fracture strain, ϵ_c, corresponds to fibre breakage and the resulting interfacial debonding. This region is hatched by vertical lines to indicate that the scatter here is on the fracture strain (see subsequent discussion). The next region is shown as a sloping band between the lower bound of the fibre-breakage scatter band and the horizontal line representing the fatigue limit of the matrix, ϵ_m. This region corresponds to matrix cracking and interfacial shear failure. These two mechanisms are taken together in one band as they may occur simultaneously, particularly towards the end of fatigue life.

The fibre breakage band in the fatigue-life diagram represents the non-progressive part of the fatigue damage. The non-progressive nature of the fibre breakage is illustrated in Figure 3. On first application of the maximum stress (strain) the fibres having lower strength than the applied stress break. The sites of fibre breaks are scattered randomly in the volume of the composite.

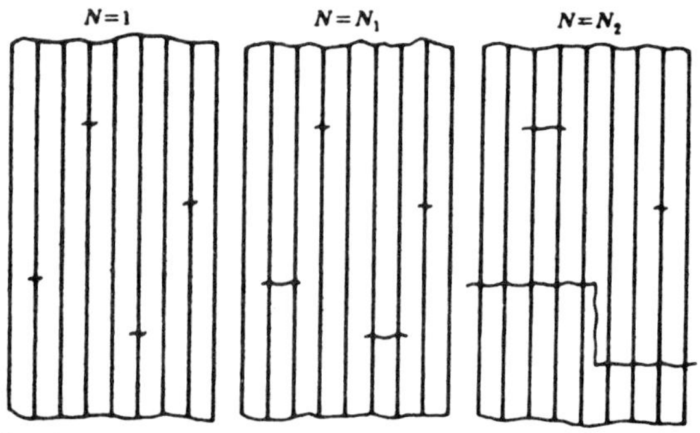

Figure 3. Fibre breakage in unidirectional composites under loading parallel to fibres.

Repeated application of the stress results in breakage of the fibres adjacent to the previously broken fibres, due to the stress concentration in the vicinity of the broken fibres. The number of broken fibres increases with the number of cycles applied, while the distribution of the broken-fibre sites remains random. With the increasing number of broken fibres, the probability increases of finding a cross section having a stress high enough to break the remaining fibres in it. The final fracture may extend over more than one cross section.

The process of fibre breakage described is non-progressive in the sense that a zone of damage growing from an early stage of the fatigue life to the final failure cannot be traced. As a consequence, the scatter in the fibre breakage band is not dependent on the fatigue life. The apparent fatigue-life dependency in this band results from the fact that the probability of finding a cross section with enough broken fibres to cause failure under the applied maximum strain increases with increasing number of cycles. The non-progressive nature of damage in this region is reflected in the lack of degradation observed by Awerbuch & Hahn (1977) in their strength data. A slight increase in strength may be found initially, as the elimination of the weakest fibres would increase the average strength of the remaining fibres (Awerbuch & Hahn 1977).

The progressive damage occurs at strains below the scatter band of the fracture strain (i.e. strength for linear elastic composites). This damage is cycle-dependent and its rate is strain-dependent. If it is assumed that damage progresses randomly, then its cycle-dependency would lead to a scatter in fatigue life. If it is assumed that damage rate depends on the maximum strain, then an increasing fatigue life is obtained with a decreasing strain.

As the progressive damage is matrix-dependent, its initiation would depend on the crack initiation in the matrix material. The cyclic maximum strain below which no cracks or only non-propagating cracks may be initiated in the matrix material, defined as the fatigue limit of the matrix, may be obtained by fatigue testing the unconstrained (i.e. unreinforced) matrix material. This matrix material property, denoted by ϵ_m, is taken as the lower limit of the progressive matrix damage.

EFFECT OF COMPOSITE STIFFNESS ON FATIGUE-LIFE DIAGRAMS

As discussed, the fatigue limit of a matrix is a material property and will therefore be fixed for a given matrix material. The fracture strain of the composite, ϵ_c, will, however, depend on the fibre stiffness. For low stiffness fibre composites, e.g. glass-epoxy, a characteristic set of stress-strain curves is shown in Figure 4a, and for high stiffness fibre composites, e.g. graphite-epoxy, Figure 4b shows the characteristic stress-strain curves. As seen in these figures, the fracture strain ϵ_c is much higher than ϵ_m for low stiffness composites. For high-stiffness composites the difference is not as great; indeed, ϵ_c may then be even less than ϵ_m as we shall see later.

The fatigue-life diagram for glass-epoxy of three different volume fractions, V_f, is shown in Figure 5. Figures 6–8 show the fatigue-life diagrams for graphite-epoxy composites with three different fibre types. The data shown in these diagrams have been converted to strain, whenever they were reported in terms of stress, by using the relevant E-moduli. The scatter band about the mean fracture strain for glass-epoxy was estimated from the reported data. For graphite-epoxy the scatter band represents probabilities of failure of 0.05–0.95, estimated by assuming a two-parameter Weibull distribution with a shape-parameter value of 24, as reported by Hashin & Rotem (1973), except for the data in Figure 7 where the reported shape parameter was 7.77 (Awerbuch & Hahn 1977). The fatigue limit of epoxy is taken as 0.6% strain based on the fatigue-life data for epoxy reported by Dharan (1975).

Several interesting features can be observed in Figures 5–8. Most of the data in Figure 5 fall in the matrix and interface-damage scatter band. This is characteristic for low stiffness composites where the matrix is strained beyond its fatigue strain limit. Figures 6–8 illustrate the effect of fibre stiffness on the composite fatigue properties. Figure 6 shows data for high modulus graphite-epoxy that has fracture strain less than the fatigue strain limit of epoxy. As seen here, all the data fall in the fibre breakage scatter band. A true fatigue process (i.e. progressive damage) does not occur in this composite.

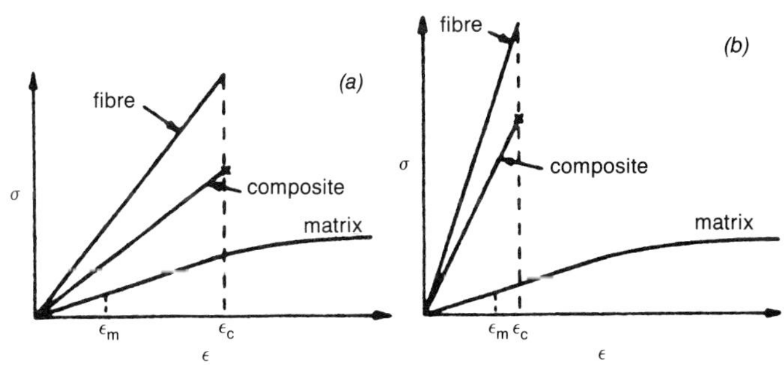

Figure 4. Stress-strain characteristics of unidirectional composites: (a) low stiffness fibres; (b) high stiffness fibres.

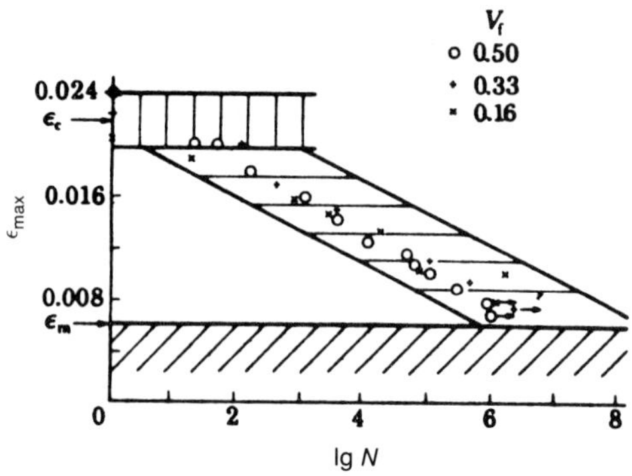

Figure 5. Fatigue-life diagram for glass-epoxy under loading parallel to fibres; data from Dharan (1975).

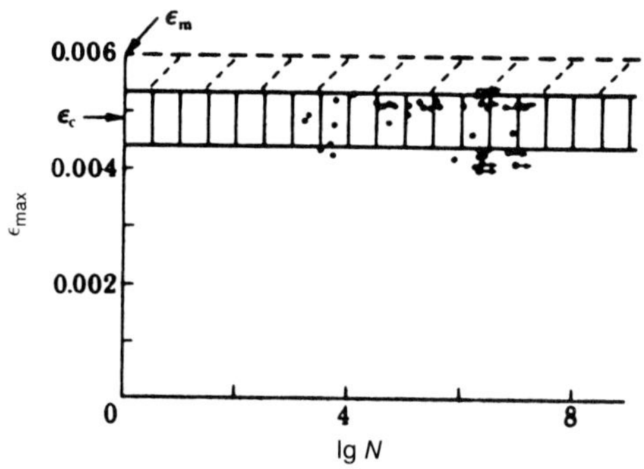

Figure 6. Fatigue-life diagram for type-I graphite-epoxy under loading parallel to fibres; data from Sturgeon (1973).

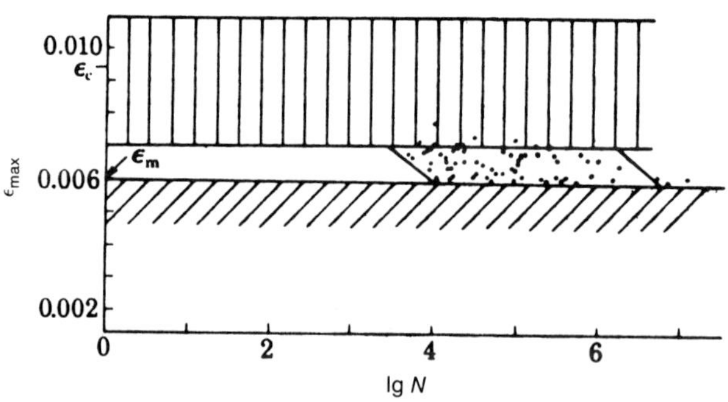

Figure 7. Fatigue-life diagram for type-II graphite-epoxy under loading parallel to fibres; data from Awerbuch & Hahn (1977).

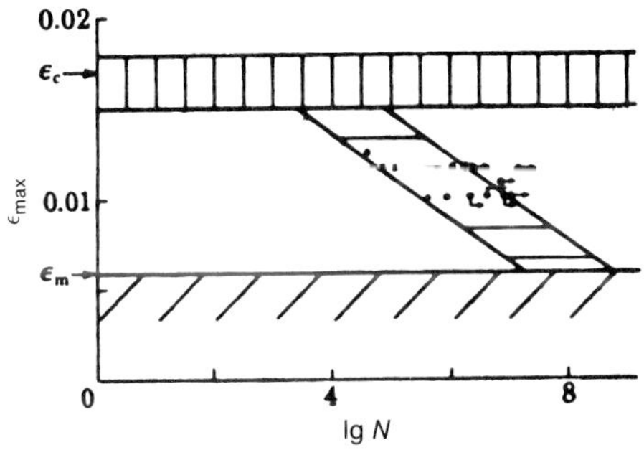

Figure 8. Fatigue-life diagram for type-III graphite-epoxy under loading parallel to fibres; data from Sturgeon (1975).

Figure 7 shows data for a type-II graphite-epoxy composite. The matrix fatigue strain limit falls slightly below the lower bound of the fibre-breakage scatter band. Progressive damage is therefore possible in a small strain range for this composite. The data in Figure 8 are for type-III graphite-epoxy, which has low stiffness fibres. The fatigue behaviour of this composite is similar to that of glass-epoxy.

FATIGUE UNDER LOADING INCLINED TO FIBRES (OFF-AXIS FATIGUE)

The predominant damage mechanisms under off-axis fatigue are illustrated in Figure 9. For off-axis angles between 0° and 90°, the tip of a crack initiated in the matrix will be subject to two displacement components: an opening normal to the fibres and a sliding parallel to the fibres. This will lead to a mixed-mode crack growth parallel to the fibres (Figure 9a). The limiting value of the crack-tip displacement, under which no growth can occur, will depend on the off-axis angle. For a given applied strain, the opening component of the crack-

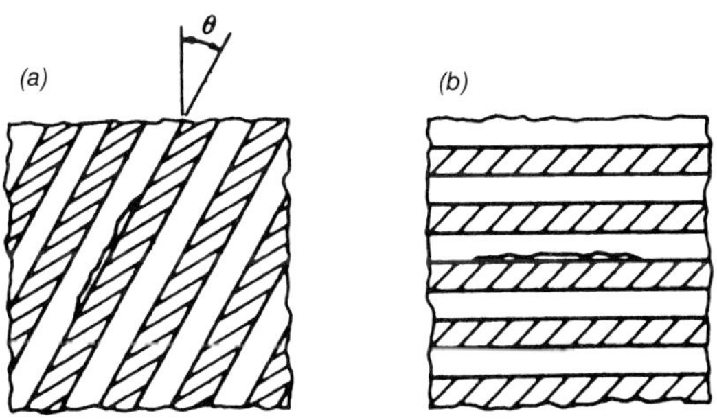

Figure 9. Matrix and interfacial cracking under off-axis fatigue of unidirectional composites: (a) mixed (opening and sliding) mode crack growth, $0 < \theta < 90°$ (b) opening mode crack growth (transverse fibre debonding), $\theta = 90°$.

tip displacement will increase with increasing off-axis angle. Now, since the opening mode of crack growth is the more critical of the two modes, the limiting applied strain, below which no crack growth can occur, i.e. the fatigue limit, will decrease with increasing off-axis angle. At an off-axis angle of 90°, i.e. when the applied loading is normal to the fibres, the crack growth will occur in the opening mode only (Figure 9b). This fracture mode will lead to debonding of the interface, referred to commonly as "transverse fibre debonding" in the literature (Owen 1974).

The fatigue-life diagram for off-axis loading is sketched in Figure 10. For completeness, the fatigue-life diagram for on-axis fatigue ($\theta = 0°$) is also sketched (dotted lines). For off-axis angles more than a few degrees, the fibre breakage band will be lost, as the matrix and interfacial cracking will be the predominant damage mechanisms for strains up to the fracture strain. The sloping scatter band will have a lower strain limit, which will depend on the off-axis angle. For the limiting case of $\theta = 90°$, the minimum strain required for transverse fibre debonding, denoted $\epsilon_{d.b.}$ here, can be expected to be a matrix material property.

A systematic investigation of the off-axis fatigue was made by Hashin & Rotem (1973). They subjected a unidirectional glass-epoxy composite to cyclic loading at various off-axis angles between 0° and 60°. Their fatigue-life data, reported in terms of stress, were converted to strain by transforming the longitudinal E-modulus with the use of the elastic constants reported by them (Rotem & Hashin 1976). Figure 11 shows these data for off-axis angles of 5°, 10°, 30° and 60°. The fracture strains at these angles are plotted at $N = 1$. The fatigue-life curves are fitted by eye to the data and are made to start at

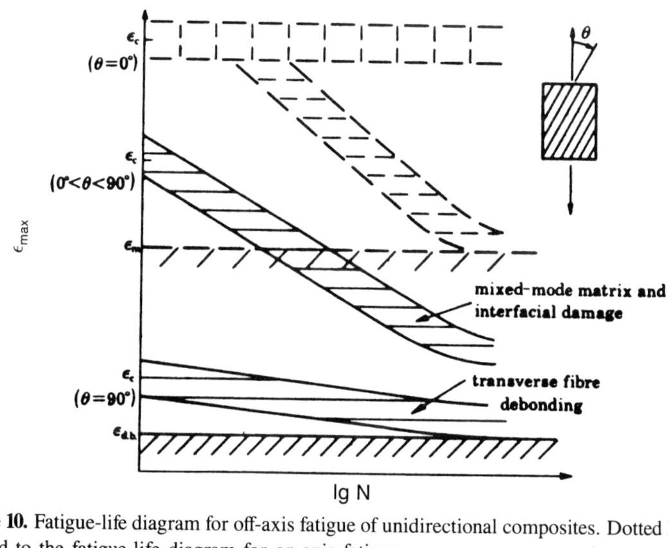

Figure 10. Fatigue-life diagram for off-axis fatigue of unidirectional composites. Dotted lines correspond to the fatigue-life diagram for on-axis fatigue.

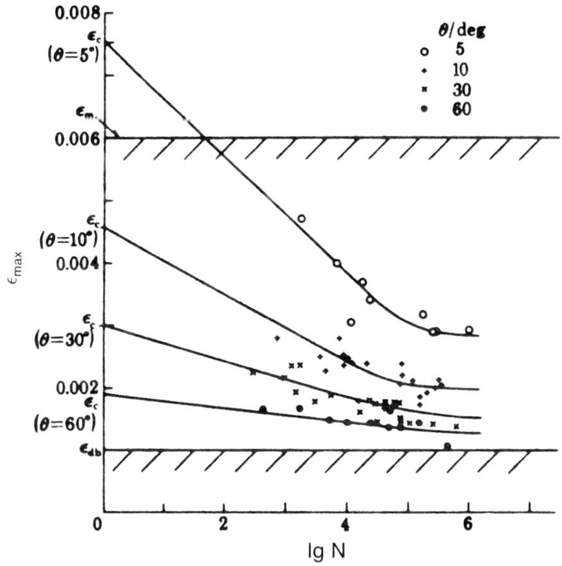

Figure 11. Fatigue-life diagram for off-axis fatigue of a unidirectional composite of glass-epoxy; data from Hashin & Rotem (1973).

the respective fracture strains. The fatigue limit of the unreinforced epoxy, ϵ_m, taken from the data by Dharan (1975) is plotted here for comparison.

At large off-axis angles, the differences in the limiting fatigue strains appear to reduce. The data suggest that the limiting strain for $\theta = 90°$ will lie close to 0.1%. This value is taken as the limiting strain for transverse fibre debonding, $\epsilon_{d.b.}$. It is interesting to note here that, on plotting the number of cycles for the onset of debonding against the maximum cyclic strain for various short-fibre composites and a cross-plied composite, it was found that a limiting strain for debonding existed at about 0.12% (Owen 1974).

In Figure 12 the limiting fatigue strain, $\epsilon_{f.l.}$, is plotted for all the off-axis angles for which data were reported by Hashin & Rotem (1973). A smooth curve is drawn through these points, and the end points taken as $\epsilon_m = 0.006$ at $\theta = 0°$ and $\epsilon_{d.b.} = 0.001$ at $\theta = 90°$.

Angle-Plied Laminates

The damage mechanisms in angle-plied laminates are basically those observed in off-axis fatigue of unidirectional composites, with the added feature of delamination. The delamination is caused by the interlaminar stresses active in the matrix layer between the off-axis plies. Thus, delamination is a matrix damage mechanism and will therefore be governed by the matrix properties.

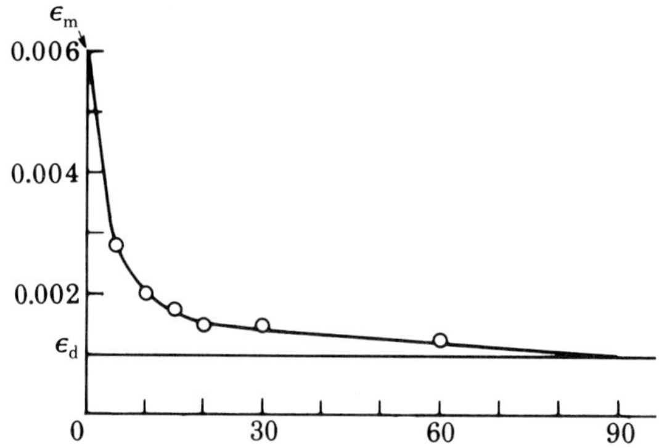

Figure 12. Variation of the fatigue limit with the off-axis angle.

Fatigue-life data for symmetric angle-plied laminates of glass-epoxy with fibre orientations between $\pm 30°$ and $\pm 60°$ were reported by Rotem & Hashin (1976). From their fatigue curves, the limiting fatigue strains were taken and are plotted in Figure 13. As seen here, the variation of the limiting fatigue strain suggests an upper bound and a lower bound. The upper bound is close to the fatigue limit of the unreinforced epoxy (0.6%) reported earlier, and the lower bound is the minimum strain for transverse fibre debonding (0.1%).

For comparison between the off-axis fatigue of the unidirectional composites and fatigue of the angle-plied laminates, the curve showing the variation of the fatigue limit with the off-axis angle (Figure 12) is drawn as a dotted line in Figure 13. It can be seen that the two curves in Figure 13 merge at angles

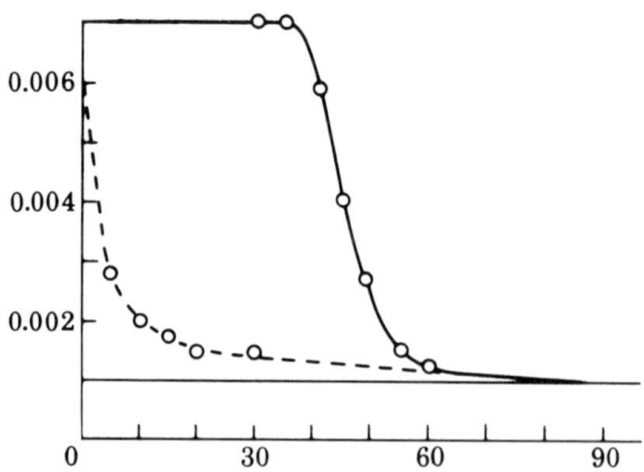

Figure 13. Variation of the fatigue limit with the fibre angle in symmetric angle-plied laminates of glass-epoxy. Fatigue limits are taken from the data reported by Rotem & Hashin (1976). The dotted line corresponds to the fatigue limit variation for off-axis angle (reproduced from Figure 12).

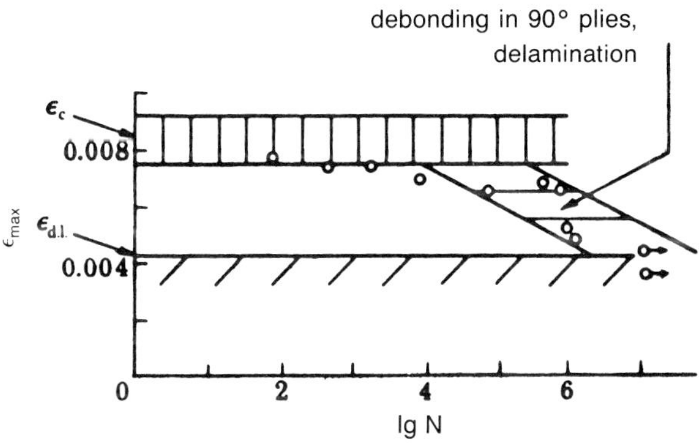

Figure 14. Fatigue-life diagram for cross-plied laminates of graphite-epoxy; data from Grimes (1977).

greater than 60°. For smaller angles, a significant improvement is obtained in the fatigue behaviour of angle-plied laminates over the off-axis fatigue of the unidirectional composites.

Cross-Plied Laminates

Investigations of fatigue damage mechanisms in cross-plied laminates by Broutman & Sahu (1969) have shown that the first event of failure is the debonding of the transverse fibres. The debond crack then grows towards the ply interfaces and on meeting them causes stress concentrations in the interfacial layers. Delamination then follows, reducing the laminate to essentially a unidirectional lamina.

A fatigue-life diagram for cross-plied laminates is shown in Figure 14 with data plotted for a graphite-epoxy laminate. The fibre-breakage scatter band exists owing to the controlling of the static strength by the 0° fibres. The progressive damage in the 90° plies leading to delamination is primarily responsible for the sloping scatter band. The fatigue limit was found to correspond to the proportional limit in the longitudinal stress-transverse-strain plot reported by Grimes (1977). The longitudinal strain at this limit was about 0.43%. This is plotted as the limiting strain for delamination, $\epsilon_{d.l.}$, in Figure 14.

Other Laminates

The most common type of laminates consists of combinations of 0°, 45° and 90° plies that are loaded along the 0° plies. The sequence of damage

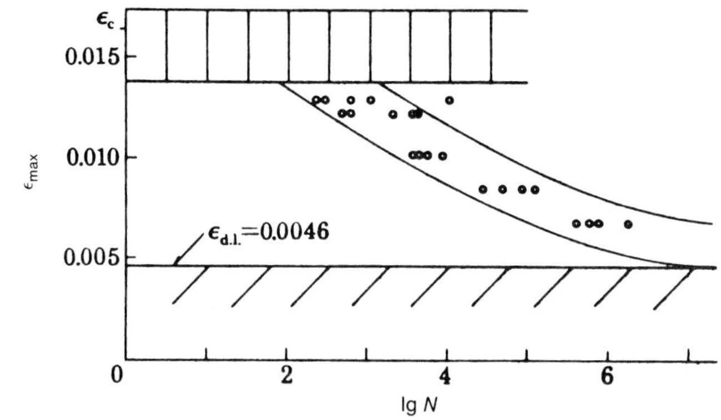

Figure 15. Fatigue-life diagram for a $(0, \pm 45, 90)_s$ glass-epoxy laminate; data from Hahn & Kim (1976).

mechanisms is found to be the failure of the 90° plies (transverse fibre debonding) followed by delaminations of the 45° plies, leading to over-stressing of the 0° plies and consequent failure.

The fatigue-life diagrams for these laminates will be expected to display the features shown by the unidirectional, the angle-plied and the cross-plied laminates. The upper part of the diagram will have the fibre-breakage scatter band followed by a sloping scatter band on fatigue life representing matrix cracking, interfacial failure and delaminations. The lower limit to the diagram will be given by the minimum strain for delamination caused by the debonding of the 90° plies. This value was found to be 0.43% for the epoxy matrix, as shown in Figure 14.

As examples of fatigue-life diagrams we show data for two laminates in

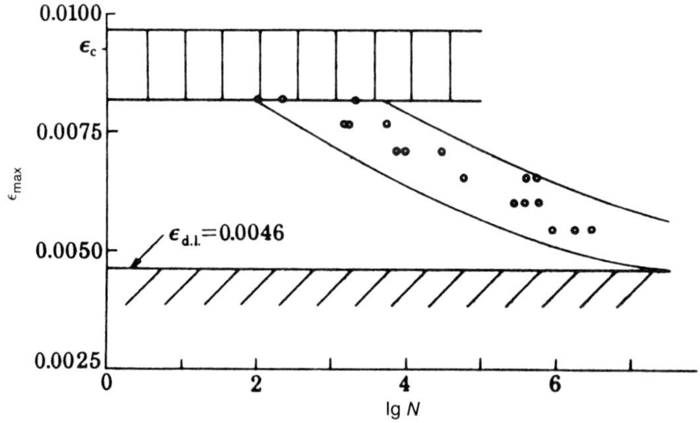

Figure 16. Fatigue-life diagram for a $(0,45,90,-45_2,90,45,0)_s$ graphite-epoxy laminate; data from Ryder & Walker (1977).

Table 1. Fatigue limits for different damage mechanisms for epoxy.

Limiting Strain in Fatigue, $\epsilon_{f.l.}$	Damage Mechanism
0.006	Matrix Cracking
0.001	Transverse Fibre Debonding
0.0046	Delamination Caused by Debonding in the 90° Plies

Figures 15 and 16. Figure 15 shows fatigue data for a $(0, \pm 45, 90)_s$ glass-epoxy laminate. The features described above are seen here and the fatigue limit converted to strain is found to be 0.46%. Figure 16 shows data for a graphite-epoxy laminate of $(0, 45, 90, -45_2, 90, 45, 0)_s$ orientation. The same features are also seen in this fatigue diagram and the fatigue limit converted to strain is also found to be 0.46%.

Fatigue Ratio and Fatigue Limit

Fatigue ratio has been defined in the literature as the ratio of the fatigue limit (in terms of stress) to the static fracture strength. From the numerous data presented here in the form of fatigue-life diagrams, it is clear that a proper representation of fatigue behaviour emerges in terms of the maximum cyclic strain. A proper definition of the fatigue ratio would therefore be as follows.

$$F = \frac{\epsilon_{f.l.}}{\epsilon_c},$$

where $\epsilon_{f.l.}$ is the fatigue limit in terms of strain and ϵ_c is the composite fracture strain. The fatigue-life diagrams reported here strongly suggest that, when expressed as strain, the fatigue limit is a matrix material property depending on the operative damage mechanisms. It represents the minimum strain required to initiate the lowest-energy damage mechanism, which often is matrix-controlled. For epoxy as the matrix material, the strain values obtained as limiting strains are summarized in Table 1, along with the corresponding damage mechanisms.

For composites whose fracture strain is lower than the unconstrained matrix fatigue limit, for example for unidirectional graphite-epoxy of type I (see Figure 6), the fatigue ratio cannot be defined because, as discussed earlier, a true fatigue process does not exist in these composites.

Conclusions

A basic pattern has been suggested for the fatigue-life diagrams for composite laminates. Various features of the composite fatigue behaviour emerge when data are plotted in accordance with these diagrams: for instance, the strain ranges in which fibre-controlled damage is operative against the strain ranges in which matrix and interfacial damage prevail. Furthermore, the limiting strain for the lowest-energy damage mechanism emerges as the fatigue limit and is strongly suggested by the data to be a matrix material property. The values of these limiting strains therefore remain independent of the fibre properties.

An important outcome of the fatigue-life diagram is the clarification of the limits to the fatigue resistance of composite materials. As seen in Table 1, the lowest fatigue performance would be in the loading mode that gives rise to the transverse fibre debonding. The performance will be much better when matrix cracking is the limiting damage mechanism. In any case, the limitations to the fatigue performance of composites is set by the matrix properties, unless the composite fracture strain is lower than the matrix fatigue limit.

The highest resistance to fatigue is achieved when the composite has only unidirectional fibres and the loading is tensile in the fibre direction. Any deviation from this will make the fatigue performance rely on the weaker phase, the matrix. Any improvement in fatigue performance will then be achieved by making the matrix material more fatigue resistant. The interface is weakest when stressed normal to the fibre direction. This loading mode must therefore be avoided. The fracture toughness of the interface could perhaps be improved by giving it more ductility.

Acknowledgement

The author would like to thank Professor H. T. Hahn and Professor K. L. Reifsnider for discussions on this work. The referee is also thanked for this comments, which were helpful in revising the paper.

References

1. Awerbuch, J. and Hahn, H. T., In *Fatigue of Filamentary Composite Materials, A.S.T.M. S.T.P.* no. 636, American Society for Testing and Materials, Philadelphia, pp. 248–266 (1977).
2. Broutman, L. and Sahu, S., *Proceedings of the 24th Conference of the Society of the Plastics Industry*, pp. 1–12 (1969).
3. Dew-Hughes, D. and Way, J. L., *Composites*, pp. 167–173 (1973).
4. Dharan, C. K. H., In *Fatigue of Composite Materials, A.S.T.M. S.T.P.*, no. 569, American Society for Testing and Materials, Philadelphia, pp. 171–188 (1975).
5. Grimes, G. C., In *Composite Materials: Testing and Design (Fourth Conference) A.S.T.M. S.T.P.*, no. 617, American Society for Testing and Materials, Philadelphia, pp. 106–119 (1977).
6. Hahn, H. T. and Kim, R. Y., *J. Compos. Mater.*, 10, pp. 156–180 (1976).

7. Hashin, Z. and Rotem, A., *J. Compos. Mater.*, 7, pp. 448–464 (1973).
8. Owen, M. J., In *Composite Materials*, ed. L. J. Broutman, Vol. 5, *Fracture and Fatigue*, Academic Press, New York, pp. 313–340 (1974).
9. Rotem, A. and Hashin, Z., *AIAA Jl.*, 14, pp. 868–872 (1976).
10. Ryder, J. T. and Walker, E. K., In *Fatigue of Filamentary Materials, A.S.T.M. S.T.P.*, no. 636, K. L. Reifnider and K. N. Lauraitis, eds., American Society for Testing and Materials, Philadelphia (1977).
11. Sturgeon, J. B., In *Proceedings of the 28th Annual Technical Conference of the Society of the Plastics Industry*, pp. 12–13 (1973).
12. Sturgeon, J. B., Technical report 75135, Royal Aircraft Establishment, Farnborough (1975).

CHAPTER 3

Damage Models for Fatigue of Composite Materials

Abstract: Mechanisms of damage in fibres, matrix and interface caused by tensile fatigue in unidirectional composites are described. Composites having brittle (e.g. glass and graphite) fibres in ductile (polymer) or brittle (e.g. glass and glass-ceramics) matrices are considered. A systematic scheme for evaluating the fatigue performance of composites is presented. This scheme is shown to reveal certain characteristics of the composite fatigue behaviour that are not easily seen by plotting the usual S-N curves. Based on the knowledge of the damage mechanisms, models for predicting the fatigue behaviour of composites are discussed. The models considered here are the statistical models for fibre damage and the continuum mechanics models for matrix and interfacial damage.

AN IMPORTANT OBJECTIVE of fatigue research is to predict the fatigue performance of materials in order to design damage-tolerant structures. A sound approach to achieve this objective is to first understand the mechanisms of damage inception and growth and then, with this knowledge, to develop models that can form a basis for calculations of the fatigue performance and design.

In the spirit of this approach, the present paper will attempt to assemble the elements of the present understanding of the fatigue damage mechanisms in unidirectional composites, based on observations and conjectures, into a systematic scheme for evaluating the role of various mechanisms in the total composite fatigue behaviour. Following this, a few damage models will be discussed. A model is here meant to be a mathematical system with preassigned constitutive behaviour and well-defined growth characteristics. The state of failure is defined as the attainment of a critical value by one or more parameters describing the model. The time to achieve the critical state is then the life to be predicted.

The composite systems considered are those with brittle continuous fibres (e.g. glass and graphite) in ductile (polymer) or in brittle (e.g. glass and glass-ceramic) matrices. The loading mode considered in tensile, applied parallel to the fibres.

Damage Mechanisms

MATRIX DAMAGE

The damage mechanisms in ductile (polymer) matrices and in brittle (glass and glass-ceramic) matrices will be considered separately.

Ductile (polymer) matrices. The fatigue process in polymers is similar to that in metals in the sense that it consists of two stages: crack initiation and crack growth (Dew-Hughes and Way 1973, Williams 1981). In fibre-reinforced polymers the matrix is subjected to strain-controlled fatigue due to the constraint provided by the fibres. The damage growth process may be described as follows. Cracks initiate from flaws in the matrix and grow until they strike an interface, whereupon further growth is delayed if the stress at the crack tip is insufficient to break the fibre. At sufficiently low strains the cracks may be confined to the matrix alone, increasing only in numbers (Figure 1a). This may be characterized as the *dispersed matrix failure mode*. At higher applied strains, the fibres at the crack tip may fail allowing then the matrix cracks to grow. If a crack is long enough, the shear stress at its tip may become sufficient to cause an interface to fail (Figure 1b) leading to a diversion of the crack in the fibre direction. In composites with high interfacial shear strength, the crack growth may be confined to a single plane normal to the fibres leading to a flat fracture surface (e.g. in surface treated high strength graphite fibres in epoxy matrix). A broom-like fracture surface may result in composites with low interfacial shear strength (e.g. in glass/polyester). In contrast to the non-growing cracks confined to the matrix alone, this failure mode may be termed as the *localized matrix failure mode*.

Brittle (glass and glass-ceramic) matrices. Highly brittle matrix materials like glass and glass-ceramics show practically no range of fatigue behaviour. This is apparently due to the very low fracture toughness of these materials

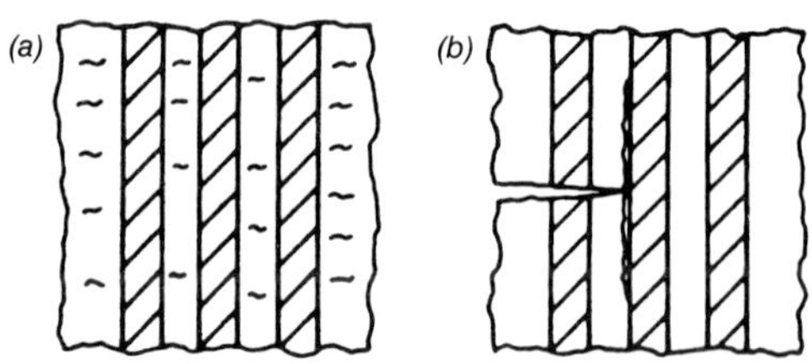

Figure 1. Ductile (polymer) matrix damage. a) Dispersed failure mode: cracks confined to matrix only. b) Localized failure mode: cracks grow by breaking fibres and, at later stage, interfaces fail.

which results in a crack of the order of an existing flaw size growing instantaneously (i.e. unstably). The negligible range of stable crack growth means that the stress (or strain) to initiate cracking may be roughly taken as the fatigue strength. The strain to initiate cracking in glass and glass-ceramics is highly influenced by the reinforcing fibres. Aveston, Cooper and Kelly (1971) put forward a theory to predict the strain for multiple fracture of brittle matrices reinforced by fibres. They show that the constraint by fibres suppresses the initiation of cracking in the matrix. The strain to cracking thus increases with the volume fraction of fibres for a given composite system.

Aveston, Cooper and Kelly assume in their work that the matrix cracks through the whole cross-section of a specimen leaving the fibres to carry all of the applied load. This is apparently true for low volume fraction of fibres. However, for high volume fraction of fibres, the obstacles to the crack growth provided by the fibres may be large enough to stop the crack, leaving the cross-section partly intact. The fibres bridging the crack may now share the additional load shed by the cracked matrix and the fibres lying in the vicinity of the crack front may be subjected to enhanced stresses caused by the crack opening. If the enhanced stress on a fibre is higher than its fracture stress, the fibre will break, leading to further growth of the crack. In fatigue, a stable growth of the matrix crack may occur around the fibres leading to, at some stage, a cross-section containing fibres only. The load transfer from matrix to fibres, the shear lag and the multiple cracking of the matrix may then occur as envisaged by Aveston, Cooper and Kelly. However, the strain required to initiate multiple cracking in fatigue may be lower than the strain calculated by Aveston, Cooper and Kelly for the constant load case.

FIBRE DAMAGE

Brittle fibres do not show a fatigue behaviour in the usual sense. To the knowledge of the present author, there has been no study that gives an evidence of a stable crack growth under fatigue for fibres such as glass and graphite. A crack growing from an existing flaw on the surface of such a fibre grows instantaneously across the cross-section. For an individual fibre, therefore, the failure may be considered as due to overstressing, i.e. static fracture.

For a bundle of fibres, such as in a composite, the fatigue damage may be seen in terms of the number of fibres broken. Since overstressing is the cause of fibre failure, a fibre may break either at the weakest point along its length or at a point of local stress concentration (such as at the tip of a matrix crack). The consequence of a single fibre failure may depend on the state of the matrix in a composite. Three situations can be imagined:

1. The matrix is intact.
2. The matrix has cracked in a dispersed failure mode.
3. The matrix has cracked fully in a cross-section, e.g. in a multiple fracture mode.

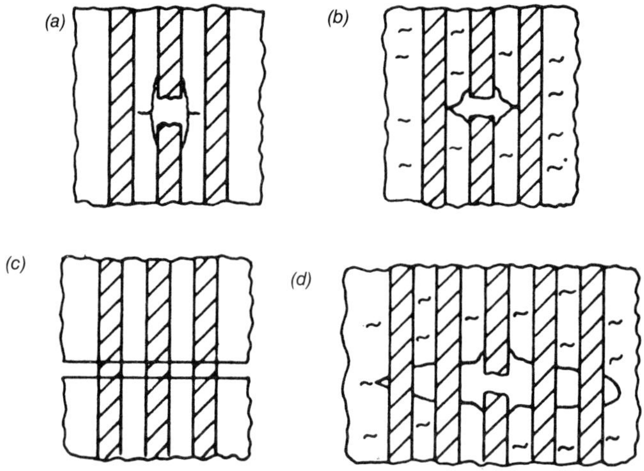

Figure 2. Fibre damage: a) fibre break causing interfacial debonding, b) fibre break increasing matrix crack, c) fibres bridging a matrix crack, and d) combination of a), b) and c).

When the matrix is intact, a single fibre failure at the weakest point in a fibre would lead to a situation illustrated in Figure 2a. The fibre failure would cause a shear stress concentration in the interface which may lead to debonding of the fibre from the matrix. The void created in the matrix would enhance the tensile stress locally leading to a transverse crack.

Figure 2b illustrates a situation where a fibre has failed near the tip of a crack in the matrix that has failed in a dispersed failure mode. The failure of a fibre here increases the crack length leading to an enhanced stress in the neighbouring fibres.

Fibre bridging of a cracked matrix is illustrated in Figure 2c. The load shed by the cracked matrix is here shared equally by the fibres bridging the matrix crack. The fibres thus behave as in a loose bundle pulled by stiff, parallel grips. The weakest fibre fails first; the load shed by the broken fibres is shared equally by the surviving fibres.

In a composite, the three situations described above may exist separately at different locations or in a combination as illustrated by Figure 2d. It is expected, however, that in a given composite the three situations may each have a region of dominance given by the range of the applied cyclic strain.

INTERFACIAL DAMAGE

An interface is not a well-defined entity. It is visualized as a thin layer consisting of the matrix and the fibre materials where the two phases are bonded together chemically or otherwise. Its strength properties are limited by the strength properties of the weaker phase.

Little is understood of the behaviour of the interface, much less its damage characteristics. As a simplistic approach, the cracking parallel to the fibres may be seen as the interfacial damage. This damage, then, has basically two modes: the tensile mode and the shear mode caused by the mode I and the mode II of crack growth, respectively. An example of the shear mode interfacial damage is shown in Figure 1b.

Fatigue-Life Diagrams

The mechanisms of damage in fibres, matrix and interface described above form a complex picture of the composite fatigue damage, when one considers that the mechanisms might act together in various combinations. However, it is plausible that the growth rates for the individual damage mechanisms might be sufficiently different such that a number of separable regions may be found in each of which a limited number of mechanisms dominate. These regions of dominance by damage mechanisms may be charted out on maps whose coordinate axes are taken as variables on which the growth rates of the mechanisms depend. Such damage mechanism maps may be useful for characterizing the fatigue behaviour of a given composite and for evaluating the relative fatigue performance of different composites. For metallic materials, fracture mechanism maps have been developed by Ashby (1977).

The primary variable on which the growth rates of the damage mechanism depends is the applied strain. We shall choose strain rather than stress as the variable since both fibres and matrix are subjected to the same strain when a composite deforms, while the stress in the two phases is in general different. As other variables, one may take the strain rate, the temperature or an environmental parameter such as the moisture content. We shall here, however, restrict ourselves to only the strain and consider the development of the damage mechanisms along the number of cycles as the axis. The critical states in the development of the damage mechanisms will be identified with the fatigue life and the diagrams showing these will be called the fatigue life diagrams.

POLYMER MATRIX COMPOSITES

A schematic fatigue life diagram for tensile fatigue of unidirectional polymer matrix composites is shown in Figure 3. The maximum applied strain is taken as the vertical axis and the logarithm of the number of cycles as the horizontal axis. The hatched areas on the diagram indicate the regions containing the critical states in the various mechanisms. The horizontal band centered about the mean composite fracture strain ϵ_c, corresponds to fibre breakage and the resulting interfacial debonding. This region is hatched by vertical lines to indicate that the scatter here is on the fracture strain. The next

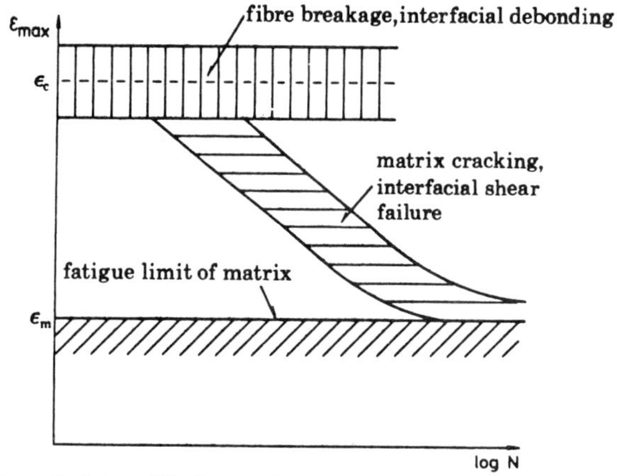

Figure 3. A schematic fatigue life diagram for tensile fatigue of unidirectional polymer matrix composites.

region is shown as a sloping band between the lower bound of the fibre-breakage scatter band and the horizontal line representing the fatigue limit of the matrix, ϵ_m. This region corresponds to matrix cracking and interfacial shear failure. These two mechanisms are together in one band as they may occur simultaneously, particularly towards the end of fatigue life.

The fibre breakage band in the fatigue life diagram represents the non-progressive part of the fatigue damage. The non-progressive nature of the fibre breakage is illustrated in Figure 4. On first application of the maximum stress (strain) the fibres, having lower strength than the applied stress, break. The sites of fibre breaks are scattered randomly in the volume of the composite. Repeated application of the stress results in breakage of the fibres adjacent to the previously broken fibres, due to the stress concentration in the vicinity of

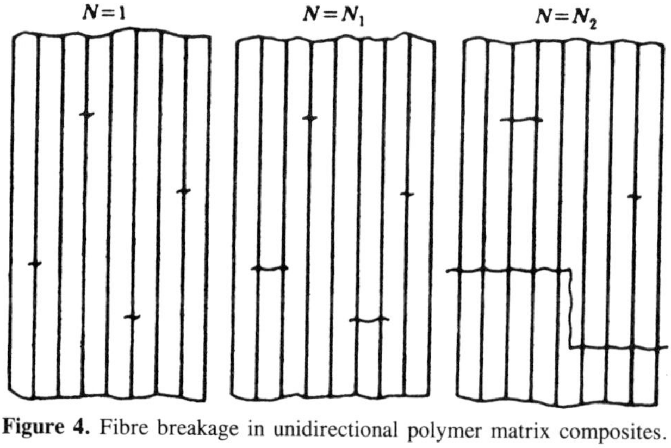

Figure 4. Fibre breakage in unidirectional polymer matrix composites.

the broken fibres. The number of broken fibres increases with the number of cycles applied, while the distribution of the broken fibre sites remains random. With increasing number of broken fibres, the probability of finding a cross-section having a stress high enough to break the remaining fibres in it, increases. The final fracture may extend over more than one cross-section.

The process of fibre breakage just described is non-progressive in the sense that a zone of damage growing from an early stage of the fatigue life to the final failure cannot be traced. As a consequence of this, the scatter in the fibre breakage band is not on the fatigue life. The apparent life-dependency in this band results from the fact that the probability of finding a cross-section with enough broken fibres to cause failure under the applied maximum strain increases with increasing number of cycles. The non-progressive nature of damage in this region is reflected in the lack of degradation observed by Awerbuch and Hahn (1977) in their strength data. A slight increase in strength may be found initially, as the elimination of the weakest fibres would increase the average strength of the remaining fibres (Awerbuch and Hahn 1977).

The progressive damage occurs at strains below the scatter band of the fracture strain (i.e. strength for linear elastic composites). This damage is cycle-dependent and its rate is strain-dependent. Assuming damage to progress randomly, its cycle-dependency would lead to a scatter in fatigue life. Assuming damage rate to depend on the maximum strain, an increasing fatigue life with a decreasing strain is obtained.

As the progressive damage is matrix dependent, its initiation would depend on the crack initiation in the matrix material. The cyclic maximum strain below which no cracks or only non-propagating cracks may be initiated in the matrix material, defined as the fatigue limit of the matrix, may be obtained by fatigue testing the unconstrained (i.e. unreinforced) matrix material. This matrix material property, denoted by ϵ_m, is taken as the lower limit of the progressive matrix damage.

As discussed above, the fatigue limit of matrix is a material property and will therefore be fixed for a given matrix material. The fracture strain of the composite, ϵ_c, will, however, depend on the fibre stiffness. For low-stiffness fibre composites, e.g. glass-epoxy, a characteristic set of stress-strain curves is shown in Figure 5a. For high-stiffness fibre composites, e.g. graphite-epoxy, Figure 5b shows the characteristic stress-strain curves. As seen in these figures, the fracture strain ϵ_c is much higher than ϵ_m for low-stiffness composites. In the case of high-stiffness composites the difference is not as great, indeed, ϵ_c may then be even less than ϵ_m as we shall see later.

The fatigue life diagram for glass-epoxy of three different volume fractions is shown in Figure 6. Figures 7–9 show the fatigue life diagrams for graphite-epoxy composites with three different fibre types. All the data shown in these diagrams have been converted to strain, whenever these were reported in terms of stress, by using the relevant E-moduli. The scatter band about the mean fracture strain for glass-epoxy was estimated from the reported data. For

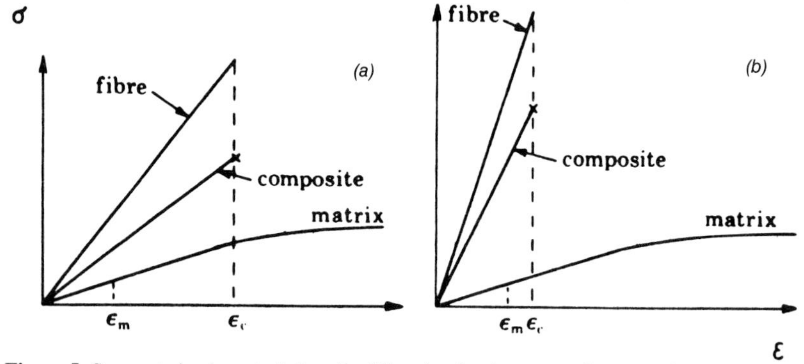

Figure 5. Stress-strain characteristics of unidirectional polymer matrix composites: a) low stiffness fibres; b) high stiffness fibres.

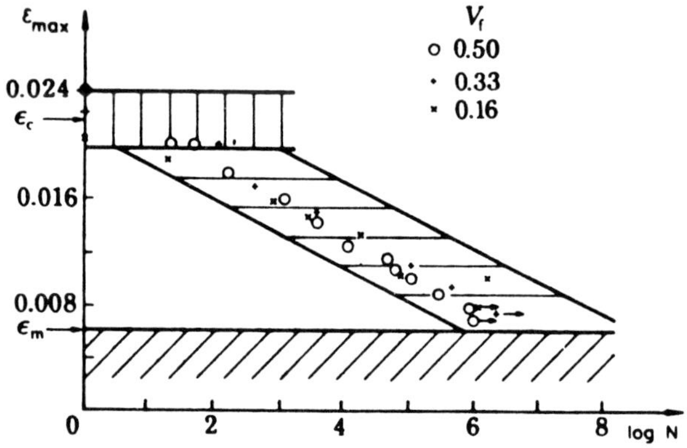

Figure 6. Fatigue life diagram for glass-epoxy unidirectional composite. Data from Dharan (1975).

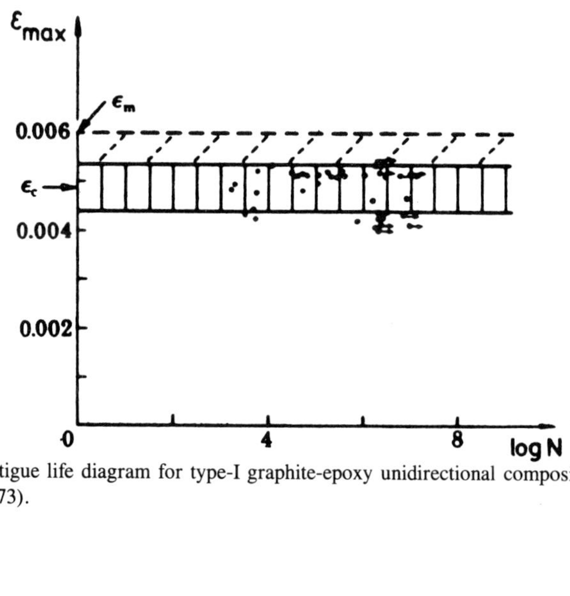

Figure 7. Fatigue life diagram for type-I graphite-epoxy unidirectional composite. Data from Sturgeon (1973).

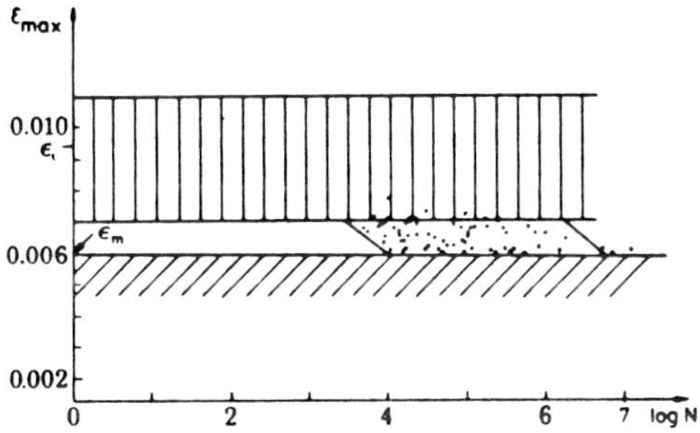

Figure 8. Fatigue life diagram for type-II graphite-epoxy unidirectional composite. Data from Awerbuch and Hahn (1977).

graphite-epoxy the scatter band represents 0.05 to 0.95 probabilities of failure estimated by assuming a two-parameter Weibull distribution with shape parameter value of 24, as reported typically in literature, except for the data in Figure 8 where the reported shape parameter was 7.77 (Awerbuch and Hahn 1977). The fatigue limit of epoxy is taken as 0.6 percent strain based on the fatigue life data for epoxy reported by Dharan (1975).

Several interesting features can be observed in Figures 6–9. Most of the data in Figure 6 falls in the matrix and interface damage scatter band. This is characteristic for low-stiffness composites where matrix is strained beyond its fatigue strain limit. Figures 7–9 illustrate the effect of fibre stiffness on the composite fatigue properties. Figure 7 shows data for high modulus graphite epoxy that has fracture strain less than the fatigue strain limit of epoxy. As seen here, all the data fall in the fibre breakage scatter band. A true fatigue process (i.e. progressive damage) does not occur in this composite. Figure 8 shows data for type II graphite-epoxy composite. The matrix fatigue

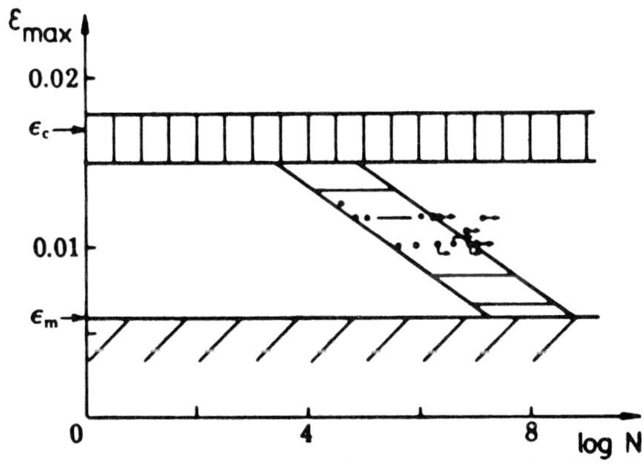

Figure 9. Fatigue life diagram for type-III graphite-epoxy unidirectional composite. Data from Sturgeon (1975).

strain limit falls slightly below the lower bound of the fibre breakage scatter band. Progressive damage is therefore possible in a small strain range for this composite. The data in Figure 9 is for type III graphite-epoxy, which has low-stiffness fibres. The fatigue behaviour of this composite is similar to that of glass-epoxy. Fatigue life diagrams for off-axis loading of unidirectional composites and for various laminates are discussed further in Chapter 2.

GLASS AND GLASS-CERAMIC MATRIX COMPOSITES

As discussed earlier, glass and glass-ceramics exhibit practically no stable crack growth. Brittle fibres of glass and graphite do not have an appreciable stable crack growth either. At the outset, therefore, it may seem that composites consisting of brittle fibres in brittle matrices do not have a significant fatigue effect. It is likely, however, that strength degradation under fatigue of fibres acting as a bundle may be affected by the manner in which the matrix fails. We shall elaborate upon this statement below.

According to Aveston, Cooper and Kelly (1971) the strain required to initiate multiple cracking in a brittle matrix reinforced by continuous fibres is given by

$$\epsilon_{muc} = \left(\frac{12\tau\gamma_m E_f v_f^2}{E_c E_m^2 r v_m} \right)^{1/3} \tag{3.1}$$

where τ is the interfacial shear strength, γ_m is the surface energy of a matrix crack and r is the fibre radius. The remaining symbols have the usual meaning.

For a given fibre (of constant diameter) and a given matrix, the strain to matrix cracking is a function of the fibre volume fraction. Figure 10 shows

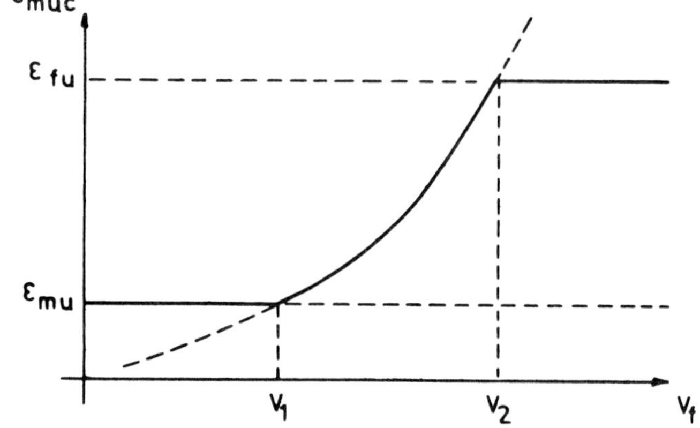

Figure 10. Strain to initiate cracking in brittle matrix composites as a function of the fibre volume fraction.

schematically the dependence of this strain on the fibre volume fraction, with the failure strain of the unreinforced matrix and the failure strain of the fibres shown as the lower and upper limits respectively (the horizontal lines in the figure). The fibre volume fraction V_1 corresponding to the point of intersection of the lower horizontal line and the curve of the strain to constrained cracking represents the limiting value of the fibre volume fraction below which cracking in a composite will initiate at the matrix failure strain. Between V_1 and V_2, the fibre volume fraction at the upper intersection point, cracking in a composite will initiate at the strain given by (3.1). For fibre volume fractions greater than V_2, a composite will fail at the failure strain of fibres.

For a composite of fibre volume fraction less than V_1, a matrix crack will grow instantaneously through the whole cross-section (normal to the fibres) exposing all fibres in that cross-section to environmental attack. The fibres here will behave as in a loose bundle of fibres that are loaded in parallel. If the fibres are susceptible to stress corrosion, a time-dependent degradation of strength of the fibre bundle will take place. Failure will occur when the stress in the surviving fibres is higher than their breaking stress. A schematic fatigue life diagram for this case is shown in Figure 11a. The fibre breakage band, initially centered about the mean fracture strain of fibres, shows degradation that increases with increasing number of cycles. The lower limit in the fatigue diagram is given by the fracture strain of the unreinforced matrix.

For composites of fibre volume fraction between V_1 and V_2, the cracking of matrix will be delayed by the constraint provided by fibres until the strain given by (3.1) is reached. However, at this strain, cracking may first initiate locally between the fibres and at the free surface (where the matrix is unconstrained). Further crack growth may be delayed by fibres acting as obstacles in the crack path. It is likely, however, that with continued cycling, a cross-section full cracked in matrix may eventually be obtained. At this stage, the fibres will behave as in a loose bundle exposed to environment. A schematic fatigue life diagram for this case is shown in Figure 11b. It may be seen that the strength degradation here is slower than in Figure 11a due to the delayed matrix cracking discussed above.

Figure 11c shows the schematic fatigue life diagram for a composite having fibre volume fraction greater than V_2. Since the matrix cracking is totally suppressed here, a composite will fail by failing the fibres first. Fibre breakage will be randomly dispersed in the volume of the composite, and failure will occur when sufficient number of fibres break in a cross-section to cause that cross-section to fail. A cycle or time-dependent strength degradation will not be expected here. This is expressed by the horizontal fibre breakage band in the fatigue life diagram. The absence of matrix cracking is indicated by the horizontal line in the diagram at the matrix cracking strain placed above the fibre breakage band.

Fatigue data for brittle matrix composites available in the literature are very

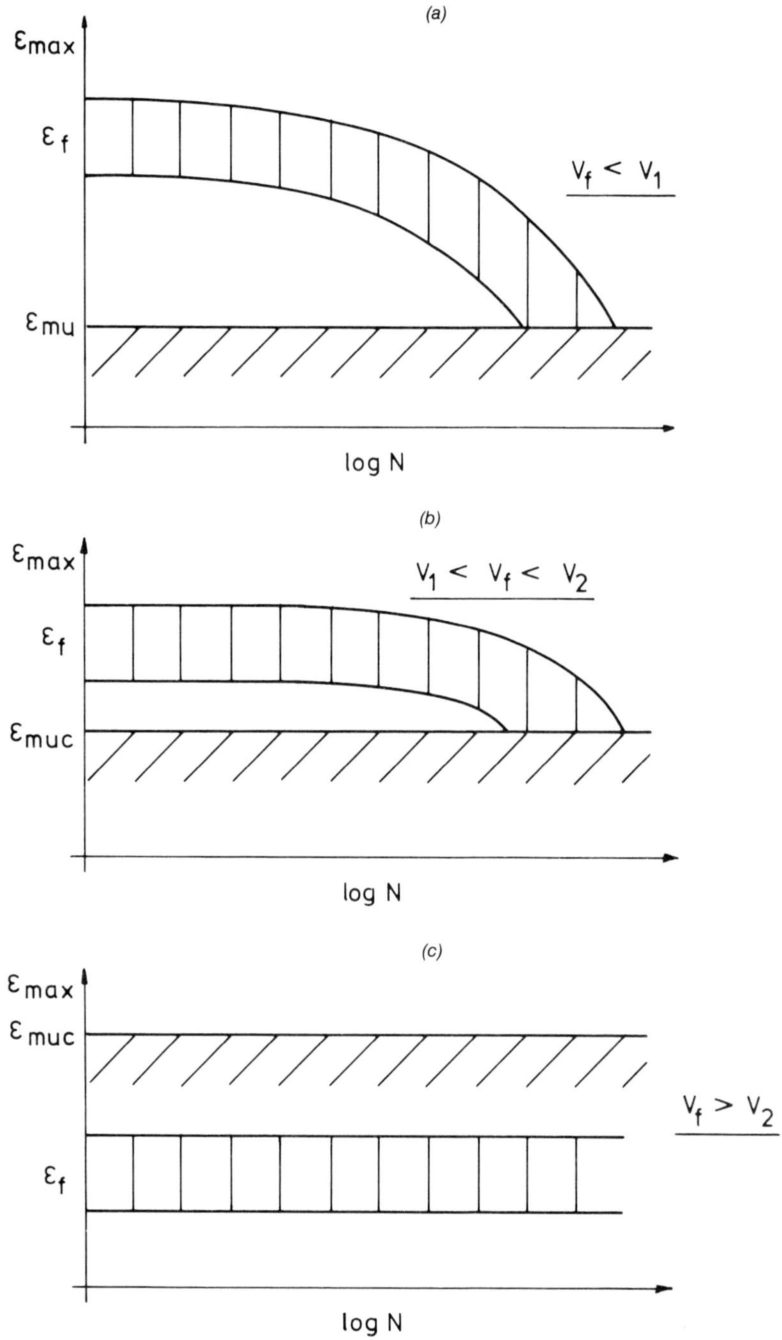

Figure 11. Fatigue-life diagram for brittle matrix composites for three ranges of fiber volume fraction.

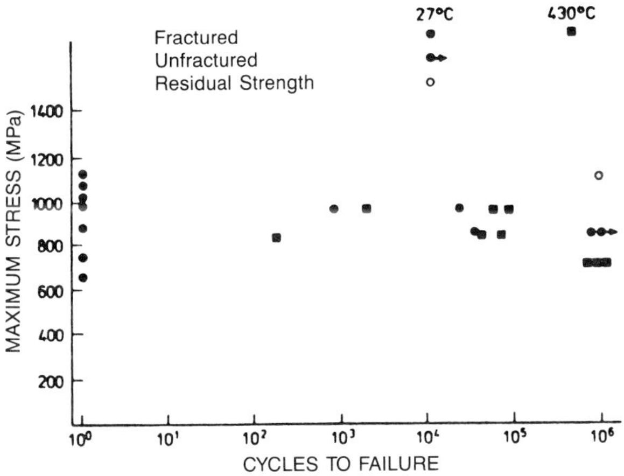

Figure 12. Flexural fatigue data for graphite fibre reinforced glass (HMS/7740M) composite.

limited. Figure 12 and Figure 13 show flexural fatigue data for two types of graphite fibres in a borosilicate glass matrix taken from Prewo and Thompson (1981). The fibre and matrix data are given in Table 1. Also given here are the strains to matrix cracking calculated by (3.1) using $\tau = 55$ MN/m^2, $\gamma_m = 4$ J/m^2 and fibre dia. of 8×10^{-6}m. As seen in the table, the strain to matrix cracking for HMS/7740 M composite is 0.606%, which is lower than the fibre failure strain of 0.77%. In accordance with the fatigue life diagram in Figure 11b, therefore, a gradual strength degradation will be expected if the fibres are susceptible to environmental damage. In Figure 12 a slight strength degradation is evident at 10^6 cycles for data obtained at 430°C, where some oxidation of fibres might take place. For GY-70/7740 M composite, the strain for matrix cracking is much higher than the fibre failure strain (0.613% versus 0.36%) indicating total suppression of matrix cracking and hence no strength degradation of fibres. In Figure 13, where data for this composite is plotted, no strength degradation is evident even for 430°C at 10^6 cycles.

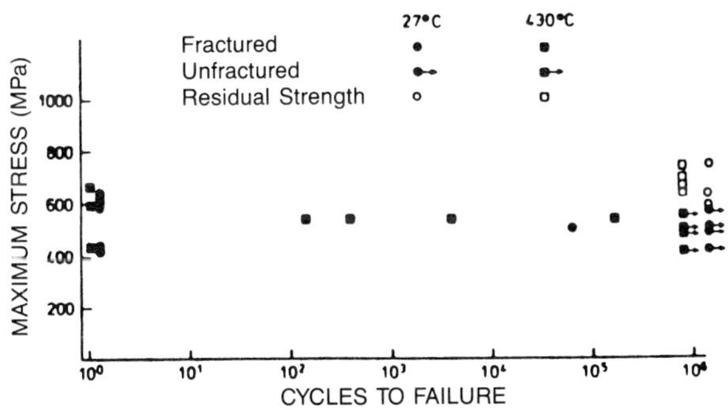

Figure 13. Flexural fatigue data for graphite fibre reinforced glass (GY-70/7740M) composite.

Table 1. Data for graphite reinforced glass composites. Elastic moduli in GPa and strain in percent.

Composite	E_f	E_m	V_f	ϵ_f	ϵ_{muc}
HMS/7740 M	350	63	0.6	0.77	0.606
GY-70/7740 M	516	63	0.6	0.36	0.613

It must be noted here that the stresses (and strains) to failure of the two composites plotted in Figure 12 and Figure 13 are much less than the failure values for the respective fibres measured without the matrix. This is apparently due to the damage incurred in the fibres during the composite fabrication process. This may mean that the strength degradation of composites under fatigue may be enhanced due to the damage already incurred during fabrication.

Damage Models

In the light of the discussion above of damage mechanisms in fibres, matrix and interace, we shall consider models or fibre damage and for matrix and interfacial damage separately.

FIBRE DAMAGE MODELS

Fibres of glass or graphite suffer no fatigue damage in the form of slow, stable crack growth. If the fibres are exposed to environment, however, their strength may deteriorate with time due to stress corrosion, oxidation etc. In a composite, a fibre may thus either break instantaneously due to overstressing or lose its strength gradually due to environmental attack. In what mode the fibre damage occurs will depend on the way the matrix behaves, as discussed in the previous section. For the purpose of predicting the composite strength due to fibre damage, two types of models are thus required. These will be discussed below.

The first type of model is the (loose) fibre bundle model illustrated by Figure 14a. The fibres are here visualized as placed parallel to one another and loaded by displacing two stiff end-plates that grip the fibres at each end. The end-plates are free to move in the fibre direction, thus allowing the fibres to extend freely. In most models the fibres are considered linearly elastic and it is assumed that, when a fibre breaks, the surviving fibres carry the additional load without preference. This so-called equal load-sharing rule has been used extensively in models to study the strength distribution of a fibre bundle (Daniels 1945, Phoenix and Taylor 1973). Recently Kelly and McCartney (1981) have investigated the effect of stress corrosion on a fibre bundle strength using the equal load-sharing rule. Their prediction of time to failure of a fibre bundle agrees with experimental results (Aveston, Kelly and Sillwood 1980).

The usefulness of the equal load-sharing rule is limited in composites to the case where a matrix crack grows through the whole cross-section exposing fibres to environmental attack, e.g. in multiple cracking of brittle matrix composites. In a case where matrix is intact, a fibre break would lead to a local failure of matrix and interface (see Section 2 above). The consequence of this is that a few fibres in the vicinity of the local failure would receive stress enhancement. The equal load-sharing rule will not express this situation correctly. Several local load-sharing rules have therefore been proposed and used in fibre bundle models. These rules define stress enhancement factors for unfailed fibres near a failed fibre.

Since a fibre failure causes a local damage, the stress enhancement on the neighbouring fibres will also be localized, i.e. limited to a small length on either side of the fibre break. To account for this, a model referred to as the "chain of bundles" model has been proposed. As illustrated by Figure 14b, this model is viewed as a large number of short bundles connected in series. Within each bundle the fibres obey a prespecified load-sharing rule. The strength of the system is given by the lowest of the strengths of the constituent bundles. The model of this type has been the basis of various studies of composite strength (Zweben and Rosen 1970, Harlow and Phoenix 1978a,b 1979, etc.).

The chain of bundles model has been used for studies of the composite strength in static loading only. In the work of Phoenix and co-workers, bounds to the probability of failure have been given by probabilities of two or more adjacent fibres failing. This has been done for the case of an increasing load where failure occurs at the load where a given number of adjacent fibres fail. In fatigue, this approach may be modified by considering the effect on the local

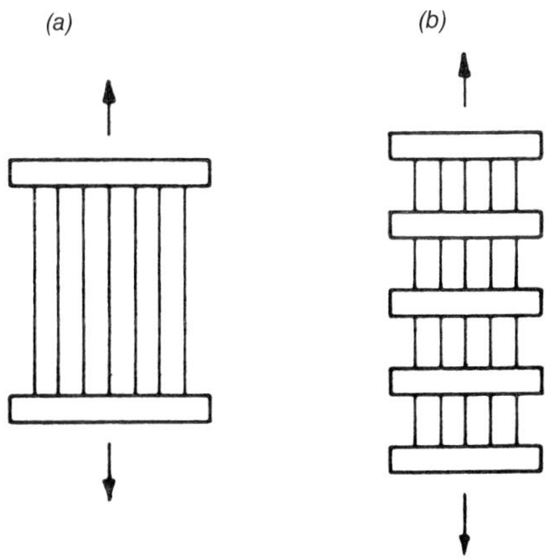

Figure 14. Fibre damage models: a) (loose) fibre bundle model, b) chain of bundles model.

load-sharing by the growth with cycles of the local failure zone caused by a fibre failure. The number of cycles of a given maximum load at which a prespecified number of adjacent fibres fail may then be taken as fatigue life.

MATRIX AND INTERFACIAL DAMAGE

As discussed in the previous section, matrix damage has two modes: a dispersed mode and a localized mode. Interfacial damage is usually localized, for instance when it results from matrix cracks normal to fibres. We shall consider the two modes separately in the following.

A dispersed matrix failure is visualized as consisting of cracks of an average size smaller than the average fibre spacing distributed throughout the volume of a composite. In fatigue, the intensity of this damage increases, which is expressed by the crack density as a function of the number of cycles. With increasing cycles, the crack density may reach a critical value, where-upon the failure may become localized.

As long as the failure mode remains dispersed, a composite may be expected to remain homogeneous. Its overall properties may however change. In particular, the deformation characteristics may be affected. The strength properties may not change until the damage becomes localized.

For the purpose of modelling the dispersed failure mode, the continuum mechanics approach is perhaps ideal. A composite may be considered a homogeneous solid whose constitutive behaviour changes with the intensity of damage. Damage may thus be introduced through a set of variables in the constitutive equations. Kachanov (1958) first introduced the void density as a damage variable for creep damage. Later on the damage variable concept has been generalized and used in various failure studies, e.g. for brittle failure (Krajcinovic and Fosenka 1981).

For isotropic elastic solids, several studies (e.g. Budiansky and O'Connell 1976) have related the elastic moduli to the cracks per unit volume. For orthotropic solids, Gottesman, Hashin and Brull (1980) derived approximate expressions for the Young's modulus and the shear modulus for the case of plane stress and cracks parallel to one of the planes of symmetry. Their expression for the Young's modulus has been used by Poursartip, Ashby and Beaumont (1982) in an investigation of damage accumulation in fatigue. These authors have defined damage as crack density and have given empirical expression to damage rates found from the measured rate of decrease of the Young's modulus.

It must be realized that a model for dispersed matrix failure can only be used to predict attainment of the critical state, which is the localization of failure. Failure of composite may occur when the localized failure zone has grown to a critical size.

For localized damage, models of the fracture mechanics type may be used. However, a fundamental requirement for the application of fracture mechanics is that the crack growth occurs in a self-similar fashion. An interfacial crack may approximately satisfy this requirement. A matrix crack which grows by breaking fibres may often not have a self-similar growth.

Conclusions

For unidirectional composites of brittle fibres in brittle or ductile matrices loaded in cyclic tension parallel to fibres, the mechanisms of damage can be classified in simple categories. Diagrams showing ranges of dominance by one or two of the mechanisms can be plotted with strain as the variable. These diagrams allow easy evaluation of fatigue behaviour of a given composite and form basis for comparing different composites.

Predictive models for fibre damage are of statistical type and have been developed primarily for static loads. Damage models for matrix damage of the dispersed type can be developed with stiffness as variables. These models do not predict failure but predict localization of damage. The localized damage can, in limited cases, be modelled by fracture mechanics models.

References

1. Ashby, M. F., "Progress in the Development of Fracture-Mechanism Maps," in *Fracture 1977*, Ed. D. M. R. Taplin, University of Waterloo Press, Waterloo, Ontario, Vol. 1, pp. 1–14 (1977).
2. Aveston, J., Cooper, G. A. and Kelly, A., "Single and Multiple Fracture," in *The Properties of Fibre Composites*, IPC Sci. and Tech. Press, Guilford, Surrey, pp. 15–26 (1971).
3. Aveston, J., Kelly, A. and Sillwood, J. M., in *Advances in Composite Materials*, Ed. A. R. Bunsell, et al., Pergamon Press, Oxford, pp. 556–568 (1980).
4. Awerbuch, J. and Hahn, H. T., "Fatigue and Proof-Testing of Unidirectional Graphite/Epoxy Composite," in *Fatigue of Filamentary Composite Materials*, ASTM STP 636, Ed. K. L. Reifsnider and K. N. Lauraitis, American Society for Testing and Materials, Philadelphia, pp. 248–266 (1977).
5. Budiansky, B. and O'Connell, R. J., "Elastic Moduli of a Cracked Solid," *Int. J. Solids Structures*, *12*, pp. 81–97 (1976).
6. Daniels, H. E., "The Statistical Theory of Strength of Bundles of Threads," *Proc. R. Soc. Lond.*, *A183*, pp. 405–435 (1945).
7. Dew-Hughes, D. and Way, J. L., "Fatigue of Fibre-Reinforced Plastics: A Review," *Composites*, pp. 167–173 (1973).
8. Dharan, C. K. H., "Fatigue Failure Mechanisms in a Unidirectionally Reinforced Composite Material," in *Fatigue of Composite Materials*, ASTM STP 569, American Society for Testing and Materials, Philadelphia, pp. 171–188 (1975).
9. Gottesman, T., Hashin, Z. and Brull, M. A., "Effective Elastic Moduli of Cracked Fiber Composites," in *Advances in Composite Materials*, Ed. A. R. Bunsell, et al., Pergamon Press, Oxford, Vol. 1, pp. 749–758 (1980).
10. Harlow, D. G. and Phoenix, S. L., "The Chain-of-Bundles Probability Model for the Strength of Fibrous Materials I: Analysis and Conjectures," *J. Comp. Mats.*, *12*, pp. 196–214 (1978a).

11. Harlow, D. J. and Phoenix, S. L., "The Chain-of-Bundles Probability Model for the Strength of Fibrous Materials II: A Numerical Study of Convergence," *J. Comp. Mats.*, *12*, pp. 314–334.
12. Harlow, D. G. and Phoenix, S. L., *Int. J. Fracture*, *15*, pp. 321–336 (1979).
13. Kachanov, L. M., "On the Creep Fracture Time," *Izv. AN SSR Otd. Tekhn. Nauk. 8*, pp. 26–31, in Russian (1958).
14. Kelly, A. and McCartney, L. N., "Failure by Stress Corrosion of Bundles of Fibres," *Proc. R. Soc. Lond. A374*, pp. 465–489 (1981).
15. Krajcinovic, D. and Fosenka, G. U., "The Continuous Damage Theory of Brittle Materials, Part 1: General Theory," *J. Appl. Mech.*, *48*, pp. 809–815 (1981).
16. Phoenix, S. L. and Taylor, H. M., "The Asymptotic Strength Distribution of a General Fiber Bundle," *Adv. Appl. Prob.*, *5*, pp. 200–216 (1973).
17. Poursartip, A., Ashby, M. F. and Beaumont, P. W. R., "Damage Accumulation During Fatigue of Composites," *Scripta Metall.*, *16*, pp. 601–606 (1982).
18. Prewo, K. M. and Thompson, E. R., *Research on Graphite Reinforced Glass Matrix Composites*, NASA CR 165711, p. 149 (1981).
19. Sturgeon, J. B., in *Proceedings of the 28th Annual Technical Conference of the Society of the Plastics Industry*, pp. 12–13 (1973).
20. Sturgeon, J. B., Technical Report 75135, Royal Aircraft Establishment, Farnborough (1975).
21. Talreja, R., "Fatigue of Composite Materials: Damage Mechanisms and Fatigue Life Diagrams," *Proc. R. Soc. Lond. A378*, pp. 461–475 (1981).
22. Williams, J. G., "Fracture Mechanics of Non-Metallic Materials," *Phil. Trans. R. Soc. Lond. A299*, pp. 59–72 (1981).
23. Zweben, C. and Rosen, B. W., *J. Mech. Phys. Solids. 18*, pp. 189–206 (1970).

CHAPTER 4

A Conceptual Framework for the Interpretation of Fatigue Damage Mechanisms in Composite Materials

Abstract: A brief account is given of the historical development of the understanding of fatigue damage mechanisms in composite materials. An analysis for interpretation of the damage mechanisms developed by the present author in two previous papers (Chapters 2 and 3) is highlighted. It is suggested that while studies of the individual damage mechanisms are vital for creating understanding of the damage process, the author's analysis provides a conceptual framework for generating a synthesis of this understanding and for evaluating the roles of the damage mechanisms in determining the fatigue life.

Key Words: Composite materials, fatigue damage, fatigue-life diagrams, damage mechanisms.

WE SHALL HEREIN review the physical aspects of damage in composite materials subjected to cyclic loads. We shall not attempt an exhaustive survey but only consider studies that have aided in illuminating the essential features of the damage mechanisms. The author's own contribution to systematizing damage mechanisms and developing fatigue life diagrams, discussed in the preceding two chapters, will be highlighted. More general reviews of the field have been made in References [3–6].

Fatigue damage mechanisms in composite materials depend in general on the constituent materials. We shall primarily consider thermosetting polymers as matrix materials and fibers of glass and graphite as reinforcements. As a special case, the newly developed class of composites of glass and glass-ceramic matrices reinforced by graphite fibers will also be considered.

Among the first investigations in the nature of damage in reinforced polymers is one by Owen and Dukes [7], who observed that the first damage mechanism in polyester reinforced by chopped strand mats of E-glass was debonding of fibers lying normal to the tensile stress. Further cycling in tension led also to debonding of other fibers. The next mechanism was initiation of cracks in resin rich areas, and this appeared to be originating from the

Reprinted, with permission from the *Journal of Composites Technology and Research*, 7, No. 1 (1985).

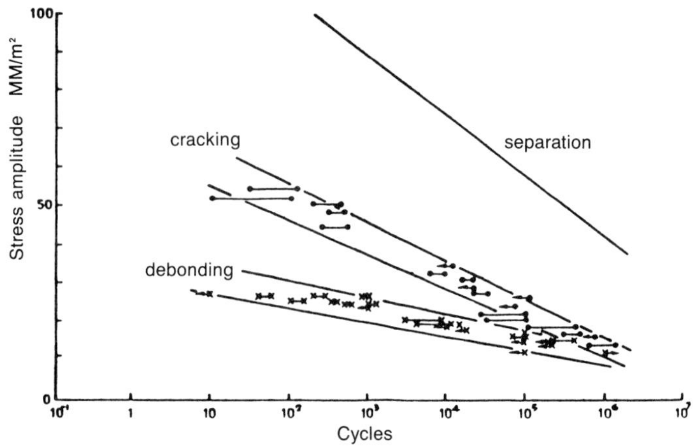

Figure 1. Fatigue life spent in initiating debonding, matrix cracking and final failure of chopped strand mats in polyester. (From Owen *et al.* [8]).

debonded regions. Owen, et al. [8] plotted the number of cycles to debonding, crack initiation and failure (separation) of specimens. The plot, reproduced in Figure 1, shows how the relative time to initiation of the two damage mechanisms changes with the applied stress amplitude. Broutman and Sahu [9] investigated cross-ply laminates of glass/epoxy and found that debonding of fibers in the transverse plies was the first damage mechanism. The debonds were found to grow into a crack spreading from one ply interface to another and subsequently spanning the width of a specimen.

Owen et al. [8] found that the maximum cyclic strain below which no debonding occurred in 10^6 cycles was about 0.12%. This strain was found to be common for chopped strand mats in polyester and cross-ply laminates of epoxy. Owen and Rose [10] investigated the possibility of improving the resistance to debonding by making the matrix more ductile. They added a flexibilizer to polyester to increase its failure strain, achieving thus a range of failure strains from 1.5% to 60%. The effect on the strain to debonding in fatigue was, however, negligible.

In a review of damage mechanisms published in 1973 Dew-Hughes and Way [11] considered damage mechanisms in the constituents, i.e. fibers, matrix and interface, and related these to the development of damage in composites. Drawing a parallel to the fatigue of metals, the authors classified the mechanisms as related to crack initiation, crack propagation and final failure. Relying on the results of Owen et al. [8], shown in Figure 1, they described debonding as the initiation mechanism which led to crack propagation in two stages, the first stage being the propagation along the interface which occurred by merger of the debonds and the second stage being the propagation through the matrix. The final failure was described as resulting from the tensile failure of the fibers. The sequential development of the fatigue damage process as perceived by Dew-Hughes and Way applies primarily to composites with low volume fraction of fibers and to short fiber composites, in the light of later work on advanced composites which has shown a more complex developmental pattern of the mechanisms involved.

A systematic and fairly thorough investigation of fatigue damage in unidirectional glass/epoxy composites was reported by Dharan [12] in 1975. He described the damage process as falling in three regions along the fatigue life axis (Figure 2). Region I, extending to 200 cycles, was found to be dominated by fiber failures caused by high stresses. The fiber failures were found to coalesce locally at different sites and grow with cycling until neighbouring zones of fiber failures joined together, weakening thereby the specimen and leading eventually to failure in a catastrophic manner. Region II covered the range from 200 cycles to 10^6 cycles. In this region the number of fiber failures was not large enough to cause failure. Instead, the matrix cracks, initiated at the surface, grew inwards by breaking the fibers at an early stage and, at a later stage, grew along the fiber/matrix interface. Region III lying beyond 10^6 cycles had no fiber failures. No matrix cracks were found initially and any cracks initiated were not found to grow until 2×10^6 cycles, at which point testing was discontinued.

At this stage it would be of interest to review the thoughts put forward regarding the rate controlling mechanism in Region II. Dharan perceived that stress corrosion cracking of glass fibers at the tip of a matrix crack controlled the cracked growth rate. However, he did not establish the validity of this suggestion. Mandel et al. [13] have recently put forward an entirely different proposal for the rate controlling mechanism. They found that the slope of the S-N curve in tension-tension fatigue normalized by the ultimate tensile strength for a variety of glass fiber composites was roughly the same and equalled the value for unimpregnated glass strands. They interpreted this as a manifestation of the strength degradation of glass fibers as the underlying cause and suggested that fiber surface friction and wear might be the mechanism involved.

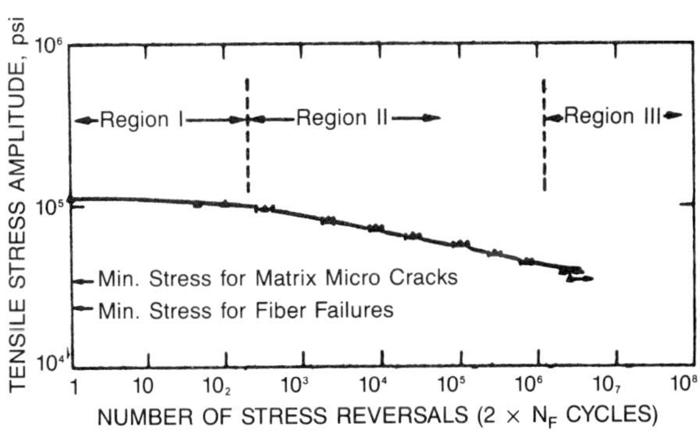

Figure 2. Three regions in fatigue life of unidirectional glass/epoxy composites under cyclic tensile loading. (From Dharan [12]).

However, an attempt to reduce friction between fibers by testing fiber strands in silicone oil did not give any measurable improvement in the fatigue degradation rate. The question regarding the rate controlling mechanism in Region II of fatigue of glass fiber composites thus remains unresolved.

Fatigue of graphite fiber composites was investigated beginning in the late 1960s when aircraft industry showed increasing interest in these materials. The early work on unidirectional composites was marred by failures initiating from the grip-ends of specimens due to the high anisotropy of graphite fibers. Also, the limited possibilities of applying microscopical techniques made investigation of damage mechanism difficult. Later investigations applied nondestructive techniques, e.g. X-ray radiography and ultrasonic scanning, and were useful in illuminating details of damage mechanisms.

Typical of the early investigations of fatigue of graphite fiber composites were those of Owen and Morris [14] and Beaumont and Harris [15]. These authors investigated effects on fatigue life of various factors, e.g. water containing environment, fiber volume fraction, fiber surface treatment, frequency of cyclic loading, etc. In most cases, unidirectional composites of high modulus graphite fiber were tested. It was found that no failures occurred in 10^7 cycles until the applied maximum stress was close to the scatter band for the ultimate tensile strength. Sturgeon [16,17] tested lower modulus graphite fiber composites and found an appreciable strength degradation in fatigue. None of the studies explained initiation and development of the underlying damage mechanisms.

In Chapter 2, I attempted a systematic characterization of tensile fatigue damage in unidirectional composites. I visualized the damage mechanisms represented by the ideal situations pictured in Figure 3 to operate when cyclic tensile loading is applied parallel to fibers in unidirectional composites. Based on these mechanisms a basic pattern for fatigue life diagrams was proposed to exist and was demonstrated, by using experimental data, to hold for unidirectional composites of glass and graphite fibers. The schematic fatigue life diagram, shown in Figure 4, has the maximum applied strain along the vertical axis and logarithm of number of cycles to failure along the horizontal axis. The scatter band centered about the composite failure strain corresponds

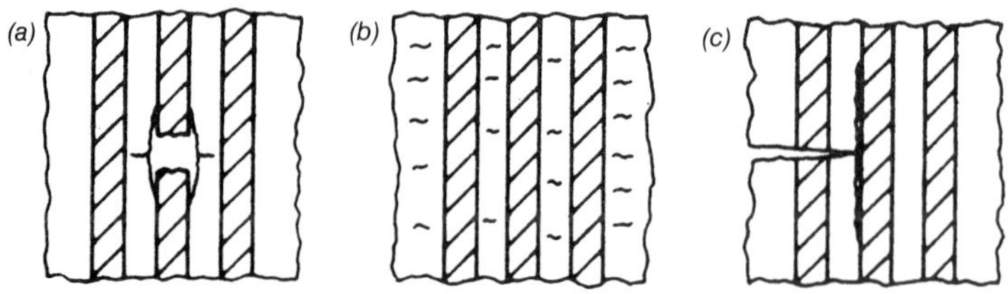

Figure 3. Fatigue damage mechanisms in unidirectional composites under loading parallel to fibers; a) fiber breakage, interfacial debonding; b) matrix cracking; c) interfacial shear failure. (From Talreja [1]).

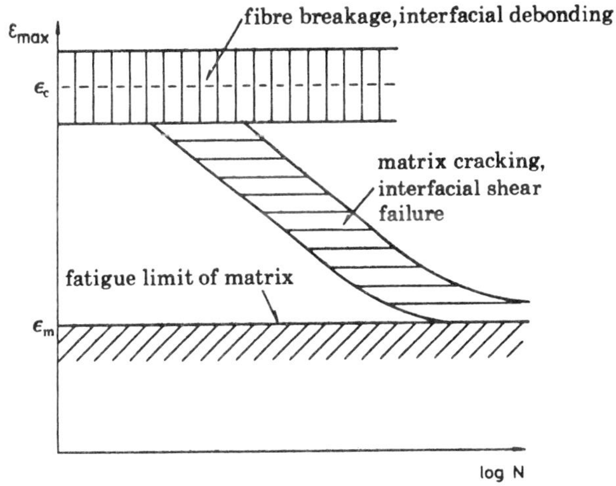

Figure 4. Fatigue life diagram for unidirectional composites under loading parallel to fibers. (From Talreja [1]).

to fiber breakage and the resulting interfacial debonding. The sloping scatter band extends from the lower bound of the fiber-breakage scatter band to the horizontal line representing the fatigue limit of matrix and corresponds to matrix cracking and interfacial shear failure.

The fiber-breakage band represents fatigue damage in unidirectional composites, which is nonprogressive in the sense that a damage zone growing from an early stage of fatigue life to final failure is not traceable. At maximum applied strain below this scatter band, the damage is progressive, leading to a strength degradation represented by the sloping band. The strength degradation rate has been demonstrated by Mandel et al. [11] to be a property of the fibers for glass fiber composites. For graphite fiber composites too, a fairly constant rate of strength degradation has been found [27]. The lower limit to the region in which progressive damage takes place, i.e. the fatigue limit, was demonstrated in Chapter 2 to be a matrix material property and was found to be about 0.6% strain for epoxy.

In Chapter 2, I demonstrated the relative roles of the progressive and the nonprogressive fatigue damage in unidirectional composites by considering composite stiffness as a parameter. In particular the fatigue life diagram was found to clarify the seemingly intricate behavior of high modulus graphite fiber composites. In these composites the failure strain falls below the fatigue limit strain of the matrix (epoxy). Thus, in accordance with the fatigue life diagram, a progressive damage is not possible and all failures are expected to fall within the scatter band for the failure strain. This also provides a likely explanation of why no appreciable effect on fatigue life of factors, e.g. loading frequency, fiber surface treatment and fiber volume fraction in high modulus

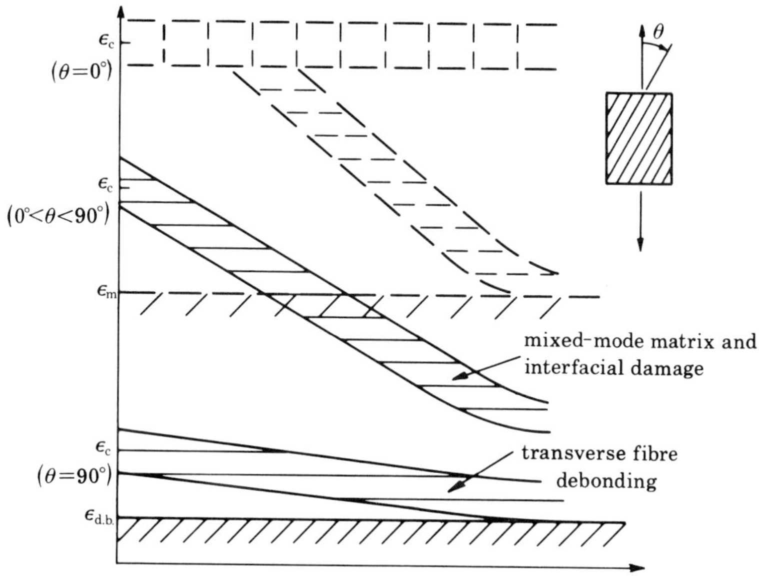

Figure 5. Fatigue life diagram for off-axis fatigue of unidirectional composites. Dotted lines correspond to fatigue life diagram for on-axis fatigue. (From Talreja [1]).

graphite fiber composites was found by Owen and Morris [14] and Beaumont and Harris [15].

Also in Chapter 2, fatigue of unidirectional composites under loading inclined to fibers (off-axis fatigue) was considered. Taking matrix cracking parallel to fibers as the operative mechanism I described the evolution of the fatigue life diagram with the off-axis angle (Figure 5). Experimental data supported the basic characteristics of the evolved diagrams for various off-axis angles. The variation of the fatigue limit strain was found to be similar to the variation of the quasi-static strength with the off-axis angle.

In Chapter 3, a systematization of the fatigue behavior of unidirectional composites with brittle matrices, e.g. glass and glass-ceramics was presented. It was proposed that the progressive damage mechanism in these composites was strength degradation of fibers which were exposed by matrix cracks to environmental attack. Thus the lower bound to the region of progressive damage, i.e. the fatigue limit, was taken to be the strain to matrix cracking.

The strain to onset of matrix cracking in composites was studied by Aveston, Cooper and Kelly [18]. Their result was utilized to classify the brittle matrix composites in three types based on the fiber volume fraction as a parameter. Thus for $V_f < V_1$, matrix cracking occurs in a single fracture mode at strain ϵ_{mu}, the failure strain of an unreinforced matrix; for $V_1 < V_f < V_2$, matrix cracks in a multiple fracture mode at a strain ϵ_{muc}, and for $V_f > V_2$, matrix does not crack before fibers fail at a strain ϵ_f. Fatigue life diagrams for the three types were proposed and are shown in Figure 6.

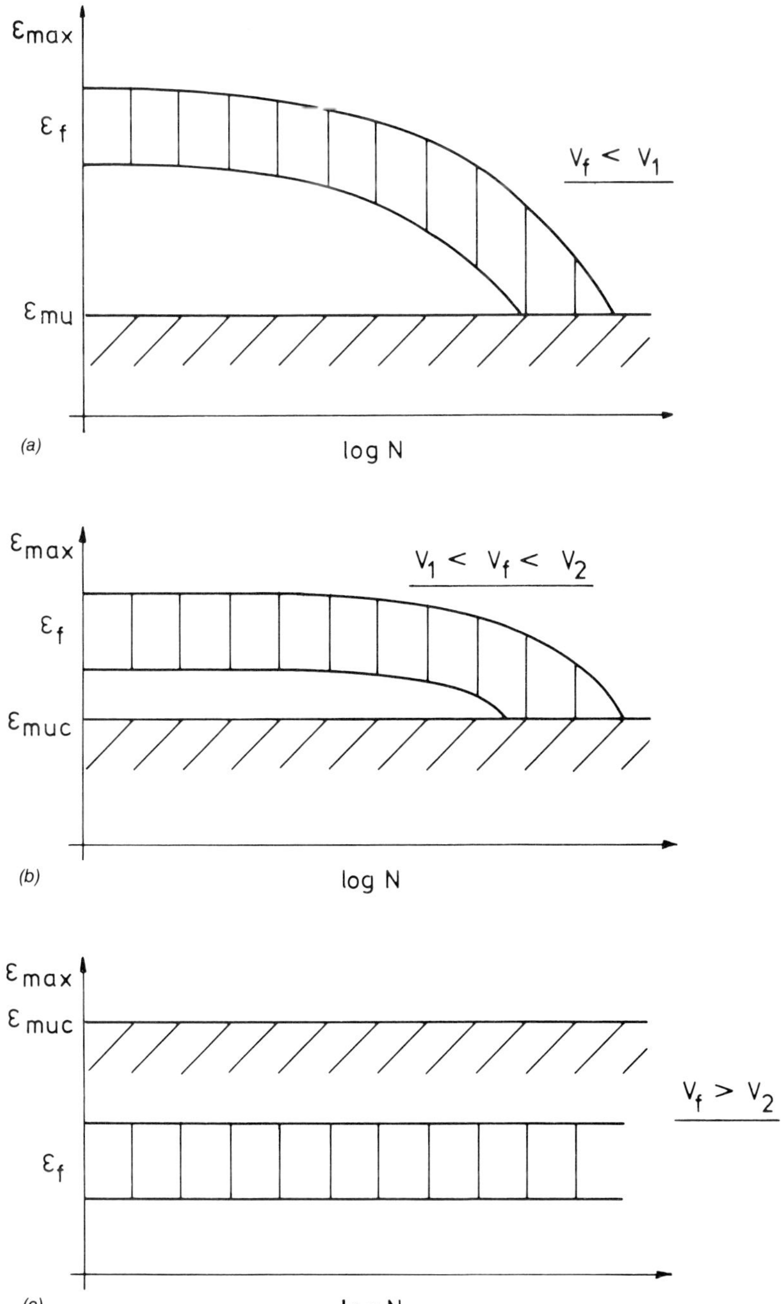

Figure 6. Fatigue life diagrams for brittle matrix composites. (From Talreja [2]).

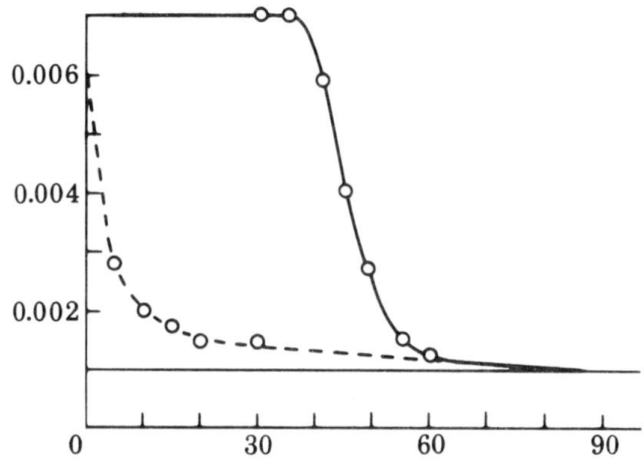

Figure 7. Variation of the fatigue limit with the fiber angle in symmetric angle ply laminates of glass/epoxy. Dotted line corresponds to the fatigue limit variation with off-axis angle in unidirectional glass/epoxy. (From Talreja [1]).

The only data available at the time on graphite fiber reinforced glass composites tested in flexural fatigue at two temperatures were examined for validity of the proposed diagrams. Although there was not a very clear trend in the data, some indication of the validity of the proposed diagrams could be found.

Studies related to unidirectional composites form useful input in analysis of fatigue behavior of laminates. For instance, it was observed in Chapter 2 that the fatigue limit in off-axis fatigue is enhanced when the off-axis plies are placed symmetrically about the loading direction. The enhancement is apparently due to constraint on matrix cracking parallel to fibers in the off-axis plies provided by the neighboring plies. The enhancement was found to be ineffective for off-axis angles greater than 60° (Figure 7).

The relative roles of fiber breakage, matrix cracking and delamination in a laminate depend on the laminate configuration. The present author in Chapter 2 demonstrated, by considering several experimental data, that fiber breakage scatter band existed whenever a 0° ply was present and that the strain to onset of matrix cracking in off-axis plies formed the lower limit to the region of progressive damage. For an epoxy matrix, I tabulated the fatigue limit strains corresponding to the matrix related damage mechanisms.

The fatigue life diagrams developed by the present author form useful means for assessing fatigue performance of a given composite material (e.g. a laminate). However, in order to predict fatigue life it is necessary not only to establish the sequential development of the damage mechanisms but also to quantify the growth rates of the mechanisms. This can only be done when the physical processes controlling the growth rates are understood in sufficient detail. In this respect the work done by the Materials Response Group at Virginia Polytechnic Institute and State University over the past several years has been of great value. Recently, Reifsnider et al. [19] have reviewed the work for the purpose of presenting a philosophy of approach to a mechanics analysis. We shall in the following draw upon this article and the works referenced in it and use the supplementary information given to the author by Reifsnider to review the physical understanding of damage mechanisms related to the rate controlling processes.

Figure 8 shows the schematics of damage development in composite laminates under fatigue as reported by Reifsnider et al. [19] for a large class of composite systems. The initial stage consists of primary matrix cracking along the fiber direction in off-axis plies. The cracking is in a multiple mode, i.e. a number of parallel cracks appear in an off-axis ply. The crack number density increases at a rate determined by the applied stress and the constraint provided by the neighbouring plies. The cracking process may continue until the cracks in each ply have attained an equilibrium or saturation spacing. The saturation spacing is a property of the laminate and is independent of the loading history. The equilibrium state in which the matrix cracking pattern is stabilized has been called the Characteristic Damage State (CDS).

As the primary matrix cracks form, fiber failure begins to initiate preferentially in adjacent plies in regions of stress concentration created by the primary cracks. In many cases, roughly two thirds of the total number of fibers that fail before fracture, fail in the first one-third of the life [20]. The primary matrix cracks, which form entirely through the thickness of the off-axis plies and generally run across the total width of a uniformly stressed region, also initiate small secondary cracks which extend short distances away from the interface between the ply in which the primary crack has occurred and the one in which the secondary crack occurs. These secondary cracks are generally

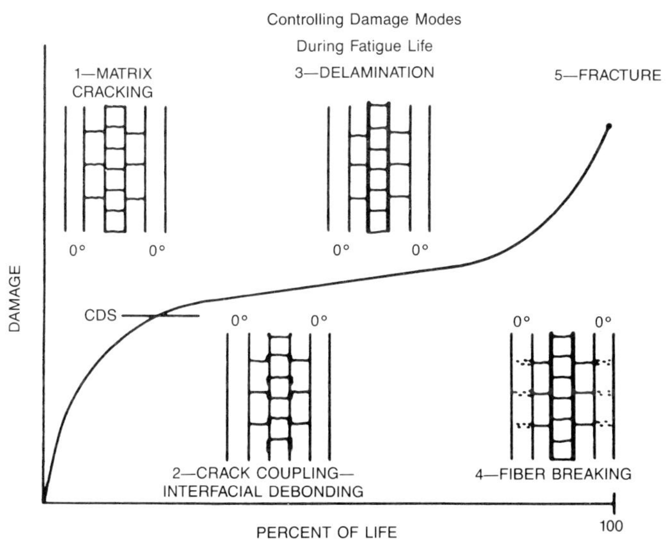

Figure 8. Development of damage in composite laminates under fatigue. (From Reifsnider *et al.* [19]).

perpendicular to the primary cracks (or as nearly so as possible) and are caused by the tensile stress along the crack axis ahead of the primary cracks.

The next event that may occur is delamination in the interior of a laminate. This delamination is caused by a mixed mode growth of interlaminar cracks which are driven by the strong interlaminar stresses in regions where primary and secondary cracks cross [21]. In compression loading the out-of-plane displacement of plies also influences the delamination growth [22].

The later stage of damage development is typified by a rapidly increasing rate of progression of all damage modes in a rapidly decreasing volume culminating in a critical local state of stress which initiates fracture of the component.

The terminal event of separation usually consists of large scale fiber failures in plies aligned with or most nearly aligned with the principal tensile stress. However, in some cases, the laminate could be considered to have functionally failed before separation due to substantial loss of stiffness.

It must be emphasized that individual damage mechanisms do not occupy distinct regions of the total life of a component but rather occur more or less simultaneously. However, for each mechanism a range of dominance may be found. For instance, primary matrix cracking may be dominant in the early stage of life while interior delamination may dominate in the intermediate stage.

Although fiber failures may occur throughout the life of a specimen, their role may not be dominant except in the terminal event of separation. However, if the applied maximum strain is large enough to be close to the average failure strain of fibers, then the fiber failure process may determine the life of the specimen. In this case, the damage development is nonprogressive and statistically controlled. The statistical analyses hitherto presented (e.g. Harlow and Phoenix [23], Batdorf [24] and Tamuzs [25]) have treated fiber failures only in unidirectional composites and have predicted that just a small number of adjacent fiber failures is needed to initiate failure of the composite. This prediction has correlated well with the experimental evidence showing that the maximum number of adjacent fiber failures after fracture of common composite systems is small, commonly of the order of 3–8 [26]. A crucial element in the statistical analyses is the stress redistribution among the neighbouring fibers on failure of a single fiber, so-called local load sharing. This may be influenced by the matrix cracking in a ply adjacent to the ply which undergoes fiber failures, as indicated by the observation that groups of broken fibers are found in bands associated with the matrix cracks [26], while a random distribution of fiber failures is found in unidirectional composites.

Conclusions

We have briefly reviewed fatigue damage mechanisms in composite materials, giving an account of the historical developments. The present understanding has resulted from a large number of investigations some of which have required developing new techniques and modifying earlier techniques for observing the complex damage process. The individual studies have illuminated parts or aspects of the process while the present author's work has attempted a synthesis to create an overall picture relating the roles played by the individual mechanisms in the process. One outcome of the synthesis has been a clarification of the limits to fatigue performance of composite materials.

The development of damage involves a sequence of mechanisms which may not be separable into distinct damage modes but rather into dominant damage modes. The parameters controlling growth rates of the individual mechanisms can only emerge from an understanding of the physical processes involved. Much progress has been made towards this objective and a major share of it is due to the work of Reifsnider and coworkers.

Acknowledgement

The author would like to thank Professor K. L. Reifsnider for his comments on this work and for providing information about the most recent work done by the Materials Response Group.

References

1. Talreja, R., "Fatigue of Composite Materials: Damage Mechanisms and Fatigue Life Diagrams," *Proc. Royal Society London*, Series A378, Chapter 2 in the present volume, pp. 461–475 (1981).
2. Talreja, R., "Damage Models for Fatigue of Composite Materials," in *Fatigue and Creep of Composite Materials, Proceedings of the Third Ris∅ International Symposium on Metallurgy and Materials Science*, H. Lilholt and R. Talreja, Ris∅ National Laboratory, Roskilde, Denmark, Chapter 3 in the present volume, pp. 137–153 (1982).
3. Salkind, M. J., "Fatigue of Composites," in *Composite Materials: Testing and Design* (Second Conference), ASTM STP 497, American Society for Testing and Materials, pp. 143–169 (1972).
4. Owen, M. J., "Fatigue Damage in Glass-Fiber-Reinforced Plastics," and "Fatigue of Carbon-Fiber-Reinforced Plastics," Chapter 7 and Chapter 8, respectively, in *Composite Materials, Vol. 5, Fracture and Fatigue*, L. J. Broutman, ed., Academic Press, New York (1974).
5. Stinchcomb, W. W. and Reifsnider, K. L., "Fatigue Damage Mechanisms in Composite Materials," in *Fatigue Mechanisms, Proc. of ASTM-NBS-NSF Symposium,*, Kansas City Missouri, May 1978, J. T. Fong, ed., ASTM STP 675, American Society for Testing and Materials, pp. 762–787 (1978).
6. Hahn, H. T., "Fatigue Behavior and Life Prediction of Composite Laminates," in *Composite Materials: Testing and Design (Fifth Conference)*, ASTM STP 674, S. W. Tsai, ed., American Society for Testing and Materials, pp. 383–417 (1979).
7. Owen, M. J. and Dukes, R., "Failure of Glass Reinforced Plastics Under Single and Repeated Loads," *Journal of Strain Analysis*, 2, No. 4, pp. 272–279 (1967).
8. Owen, M. J., Smith, T. R. and Dukes, R., *Plastics and Polymers*, 37, pp. 227–233 (1969).

9. Broutman, L. J. and Sahu, S., "Progressive Damage of a Glass Reinforced Plastic During Fatigue," *Proc. of 24th SPI Conference*, 11-D (1969).
10. Owen, M. J. and Rose, R. G., "Polyester Flexibility Versus Fatigue Behavior of RP," *Modern Plastics*, 47, No. 11, pp. 130–138 (1970).
11. Dew-Hughes, D. and Way, J. L., "Fatigue of Fibre-Reinforced Plastics: A Review," *Composites*, pp. 167–173 (July 1973).
12. Dharan, C. K. H., "Fatigue Failure Mechanisms in a Unidirectionally Reinforced Composite Materials," in *Fatigue of Composite Materials*, ASTM STP 569, American Society for Testing and Materials, pp. 171–188 (1975).
13. Mandell, J. F., Huang, D. D. and McGarry, F. J., "Tensile Fatigue Performance of Glass Fiber Dominated Composites," *Composites Technology Review*, 3, No. 3, pp. 96–102 (1981).
14. Owen, M. J. and Morris, S., "Fatigue Resistance of Carbon Fiber RP," *Modern Plastics*, p. 158 (April 1970).
15. Beaumont, P. W. R. and Harris, B., in *Proc. of Int. Conf. on Carbon Fibres, Their Composites and Applications*, Paper 49, Plastics Institute, London (1971).
16. Sturgeon, J. B., in *Proceedings of the 28th Annual Technical Conference, The Society of the Plastics Industry*, pp. 12–13 (1973).
17. Sturgeon, J. B., "Fatigue Testing of Carbon Fibre Reinforced Plastics," Technical report 75135, Royal Aircraft Establishment, England (1975).
18. Aveston, J., Cooper, G. A. and Kelly, A., "Single and Multiple Fracture," in *The Properties of Fibre Composites*, Conference Proceedings, National Physical Laboratory, IPC Science and Technology Press, pp. 15–26 (1971).
19. Reifsnider, K. L., Henneke, E. G., Stinchcomb, W. W. and Duke, J. C., "Damage Mechanics and NDE of Composite Laminates," in *Mechanics of Composite Materials, Recent Advances*, Z. Hashin and C. T. Herakovich, eds., Pergamon Press, New York, pp. 399–420 (1983).
20. Reifsnider, K. L. and Jaminson, R. D., "Fracture of Fatigue-Loaded Composite Laminates," *Int. Journal of Fatigue*, 4, No. 4, pp. 187–198 (1982).
21. Highsmith, A. L. and Reifsnider, K. L., "Internal Load Distribution Effects During Fatigue Loading of Composite Laminates," *Proc. of ASTM Symp. on Composite Materials: Fatigue and Fracture*, Dallas, Texas (October 24–25, 1984).
22. Shultz, R. W. and Reifsnider, K. L., "Fatigue of Composite Laminates Under Mixed Tension-Compression Loading," *Proc. of 2nd Int. Conference on Fatigue and Fatigue Thresholds*, (Fatigue 84), University of Birmingham, U.K. (September 1–7, 1984).
23. Harlow, D. G. and Phoenix, S. L., "Bounds on the Probability of Failure of Composite Materials," *Int. Journal of Fracture*, 15, No. 4, pp. 321–336 (1979).
24. Batdorf, S. B., "Tensile Strength of Unidirectionally Reinforced Composites—I," *J. Reinforced Plastics and Composites*, 1, pp. 153–164 (1982).
25. Tamuzs, V., "Some Peculiarities of Fracture in Heterogeneous Materials," *Proc. Second USA-USSR Symposium on Fracture of Composite Materials*, Lehigh Univ., (1981).
26. Jamison, R. D., "Advanced Fatigue Damage Development in Graphite Epoxy Laminates," Ph.D. Thesis, College of Engineering, Virginia Polytechnic Institute & State University (1982).
27. Reifsnider, K. L., private communication.

Part 2

FATIGUE DAMAGE CHARACTERIZATION

CHAPTER 5

Stiffness Based Fatigue Damage Characterisation of Fibrous Composites

Abstract: Presented is a simple procedure for measuring the changes in stiffness properties of a unidirectional fibrous composite caused by fatigue damage. Unlike previous attempts at stiffness based fatigue damage characterization, where one or two elastic constants have been investigated, we consider changes in all four independent stiffness constants of an orthotropic elastic lamina. The results obtained for a unidirectional glass fibre reinforced polyester show that while small changes occur in the longitudinal elastic modulus and the Poisson's ratio of transverse to longitudinal strains, the shear modulus and the Poisson's ratio of longitudinal to transverse strains change significantly.

The extension of this procedure to the fatigue damage characterization of laminates is discussed.

THE COMPLEXITY OF the micromechanisms of fatigue damage in fibrous composites necessitates a damage characterization based on a macroscopically measurable set of parameters. The two procedures for characterization investigated hitherto have been based on strength degradation and stiffness changes. It is generally accepted now that the strength degradation does not always reflect the fatigue damage. The stiffness properties on the other hand have been found to change continuously with fatigue cycles and therefore could provide basis for a nondestructive procedure for fatigue damage characterization. It must, however, be realized that use of this procedure assumes that the integrity of the laminate is maintained and that it behaves as a linearly elastic material.

Previous investigations related to stiffness changes, e.g. [1–4], have been content with considering a single parameter (usually the longitudinal E-modulus) to indicate the occurrence of fatigue damage. A complete characterization of fatigue damage would require measuring changes in all stiffness components and relating these to the damage mechanisms. This has

Reprinted by permission of Pergamon Press from *Advances in Composite Materials 2*, A. R. Bunsell, et. al., eds. (1980).

not been reported for any fibrous composite. The present investigation is intended as an attempt leading to the development of a procedure for a complete characterization of fibrous composite laminates.

The essential first step in the fatigue damage characterization of laminates is to measure changes in the stiffness properties of a lamina. Considering a lamina to be an orthotropic elastic material the stiffness properties are given by the longitudinal elastic modulus E_{11}, the transverse elastic modulus E_{22}, the two Poisson's ratios ν_{12} and ν_{21} and the in-plane shear modulus G_{12}. The relation $E_{11}\nu_{21} = E_{22}\nu_{12}$ reduces the number of independent elastic constants to four. The problem is now to devise an experimental procedure to record changes in all these four parameters with fatigue cycling. A continuous measurement of all four parameters without interrupting fatigue testing does not appear feasible. A procedure is therefore sought to intermittently measure the elastic constants. One approach would be to reshape the fatigue specimen, for instance by cutting out an off-axis specimen from it, at the end of a given number of fatigue cycles and measure the elastic constants on it. It would, however, be more desirable to measure all the elastic constants on the fatigue specimen and leave it undisturbed for further fatigue cycling. In a recent paper [5] a procedure to measure all four constants on a unidirectional composite, which involves tests in longitudinal, transverse, shear and flexure modes, has been proposed. Herein we present a specimen design which allows fatigue loading as well as measurement of all four elastic constants involving much less effort and time.

Material and Specimen

The material used was a unidirectional Glass Fibre Reinforced Polyester with 28% fibre volume fraction. The material was fabricated as flat plates of approximately 3 mm thickness by winding glass rovings on a rectangular frame. From the central parts of these plates the specimens were cut out and shaped.

The specimen is shown in Figure 1. The procedure for preparing the specimen was as follows. From a rectangular plate of GFRP a piece slightly larger than the specimen was cut out by a saw. A templet of steel having dimensions 1 mm less than the specimen dimensions was glued to the GFRP piece by a tape having glue on both faces. The specimen was then shaped using a rotating cylindrical steel cutter to the dimensions of the templet plus a constant distance of 1 mm. The specimen edges thus prepared were fairly flat and parallel to the templet edges, which were machined accurately to be perpendicular to the templet faces. The specimen was detached from the templet at the end of the shaping process. Aluminium tabs of 1.5 mm thickness were then glued to the specimen ends using an epoxy adhesive film. Rectangular steel plates of 6 mm thickness were connected by bolts to the spe-

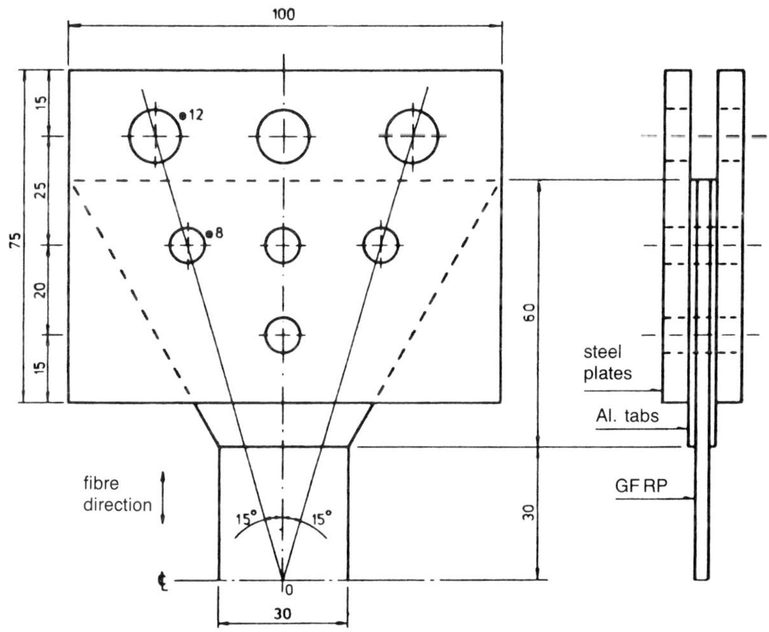

Figure 1. The specimen.

cimen ends through 4 holes of 8 mm dia. as shown in Figure 1. Strain gauge rosettes of 3 × 120° type were applied at the center of the specimen, one on each side, with one of the gauges aligned with the longitudinal axis of the specimen.

Experimental Procedure

The experimental procedure consisted of applying fatigue loads in the (longitudinal) fibre direction and stopping the test intermittently to measure the four elastic constants.

The fatigue loading was applied in a servo hydraulic testing machine at a frequency of 10 Hz. The loading was applied through pins in the 12 mm dia. holes along the longitudinal centre line in the steel plates attached to the specimen (Figure 1). The pin holders were of the type used for fracture toughness testing of compact tension specimen. Only tension-tension cycles were applied.

The procedure for measuring the elastic constants was as follows. At the end of a chosen number of cycles the specimen was unloaded to zero load. The specimen was then reloaded slowly in tension by manually operating the set point of the load signal. The analog signals of the load and the three strain outputs were plotted and the digital signals of these parameters were recorded at suitable intervals. The strain gauges of the same orientation were coupled to compensate for any bending strains. The specimen was loaded to a value below the maximum cyclic load. Records of the load and the strains were also made during unloading of the specimen. At the end of these measurements the specimen was detached by taking out the two pins from the holes along the center line. The specimen was then tilted and the two pins were placed in the 12 mm dia. holes aligned with the axis 15° off the longitudinal centre line (see Figure 1). The specimen was thus positioned to have the load axis go through its centre point and make an angle of 15° to its longitudinal axis. A tensile load was applied slowly and simultaneous records of load and strains were made at intervals. The maximum tensile load was kept below that necessary to cause shear failure parallel to fibers.

At the end of this test, which will be referred to as the 15° off-axis test, the pins were removed and the specimen tilted to place the pins in the 12 mm holes aligned with the −15° axis. A test in this position was carried out as for the 15° off-axis test.

Finally, to measure the Poisson's ratio ν_{21}, the specimen was taken out of the loading fixture and placed flat on a table. A vice was placed with its grip-plates parallel to the specimen edges and perpendicular to the plane of table. In this position the motion of the movable grip-plate was parallel to the transverse axis of the specimen. The grip-plates were brought in touch with the specimen edges and a slight clamping pressure was applied to take the initial strain readings. The specimen edges were then clamped further and the strains recorded. This was repeated a few times to check the repeatability of the strain measurements which was found to be satisfactory. Care was taken to grip the specimen edges away from the tab-ends in order to produce a uniform state of stress around the strain gauge location.

Measurement of Elastic Constants

THE LONGITUDINAL ELASTIC MODULUS E_{11}

From the longitudinal tests the records of the load vs. the longitudinal strain was taken to calculate E_{11}. Alternatively, the modulus was calculated from the off-axis tests using the following relation.

$$E_{11} = \frac{P \, Cos \, 15°}{A \, \epsilon_A} \tag{1}$$

where P is the load, A is the cross-sectional area and ϵ_A is the longitudinal strain (Figure 2).

The strain ϵ_A from the 15° off-axis test and from the −15° off-axis test should be the same for a given load. However, small differences were obtained

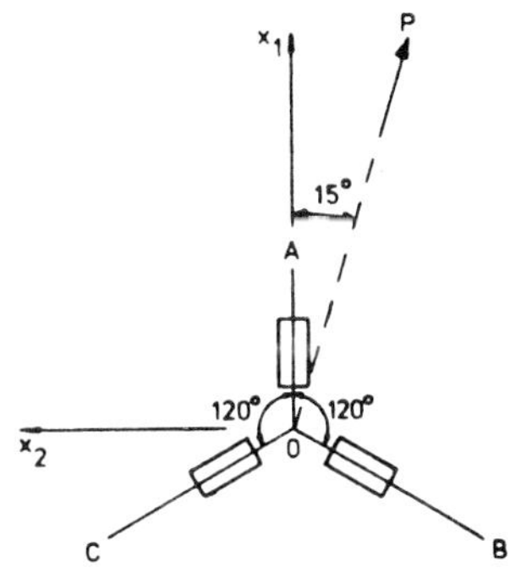

Figure 2. Orientations of the strain gauges and the load axis in off-axis test with respect to the principal material axes.

due possibly to misalignments of gauges and/or the load axis. The average of the two moduli given by the two off-axis tests was found to differ by less than 2% from the modulus calculated from the longitudinal tests.

POISSON'S RATIO ν_{12}

This ratio was calculated from the strain outputs obtained in the longitudinal tests. The following relation was used to calculate this constant.

$$\nu_{12} = -\frac{\epsilon_{22}}{\epsilon_{11}} = \frac{1}{3} - \frac{4}{3}\frac{\epsilon_B}{\epsilon_A} = \frac{1}{3} - \frac{4}{3}\frac{\epsilon_C}{\epsilon_A} \tag{2}$$

The tensor transformation of strain components is used to derive this relation.

As indicates by Equation (2), $\epsilon_B = \epsilon_C$. However, small differences in the two strains were observed due possibly to strain gauge misalignments. Average of the two strains was used in Equation (2). The ratio was calculated for five load values and almost constant value was obtained.

SHEAR MODULUS G_{12}

The shear modulus was obtained by using the load and the strain outputs from the off-axis tests. Referring to Figure 2 the shear stress is given by

$$\sigma_{12} = -\frac{P}{A}\,\mathrm{Sin}\ 15^\circ \tag{3}$$

and the shear strain by

$$\gamma_{12} = 2\epsilon_{12} = \frac{2}{\sqrt{3}}(\epsilon_B - \epsilon_C) \tag{4}$$

Thus the shear modulus is given by

$$G_{12} = \frac{\sigma_{12}}{\gamma_{12}} = \frac{\sqrt{3}}{2} \frac{P \text{ Sin } 15^\circ}{A(\epsilon_C - \epsilon_B)} \tag{5}$$

Due to the symmetry of the specimen and the strain gauges the two strains ϵ_B and ϵ_C in the off-axis test should be numerically equal and differ only by sign. However, small differences were obtained. The initial shear modulus calculated by Equation (5) was found to be in good agreement with the value obtained by an off-axis test on the same material with its fibres at 30° to the load axis.

POISSON'S RATIO ν_{21}

This ratio was calculated using the strain outputs from the transverse-clamping test using the relation:

$$\nu_{21} = -\frac{\epsilon_{11}}{\epsilon_{22}} = \frac{-3\epsilon_A}{4\epsilon_B - \epsilon_A} = \frac{-3\epsilon_A}{4\epsilon_C - \epsilon_A} \tag{6}$$

where tensor transformation of strain components is used (see Figure 2).

As indicated by Equation (6) the two strains ϵ_B and ϵ_C should be equal. Small differences were however observed and the average of the two strains was used in Equation (6). The initial ratio thus calculated was found in good agreement with the value obtained by an off-axis test on the same material with its fibres at 30° to the load axis.

Results and Discussion

Table 1 shows the measured initial values of the four elastic constants and the calculated value of E_{22}.

Figures 3–6 show the variation of the four elastic constants normalized by their initial values with the number of cycles applied. All results in these figures are for the same specimen subjected to the cyclic stress of 52 MPa ± 38 MPa.

As seen in these figures the longitudinal elastic modulus E_{11} remains essentially constant. The Poisson's ratio ν_{12} shows a slight increase. The in-plane

Table 1. The initial elastic constants.

E_{11}	E_{22}	G_{12}	ν_{12}	ν_{21}
36.110	14.598	3.157	0.235	0.095

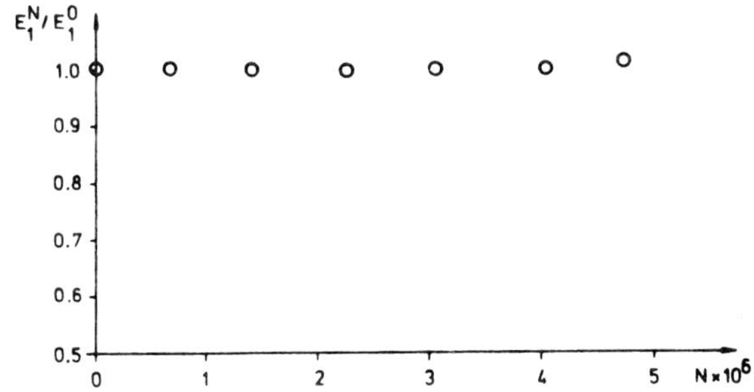

Figure 3. Variation of E_{11}.

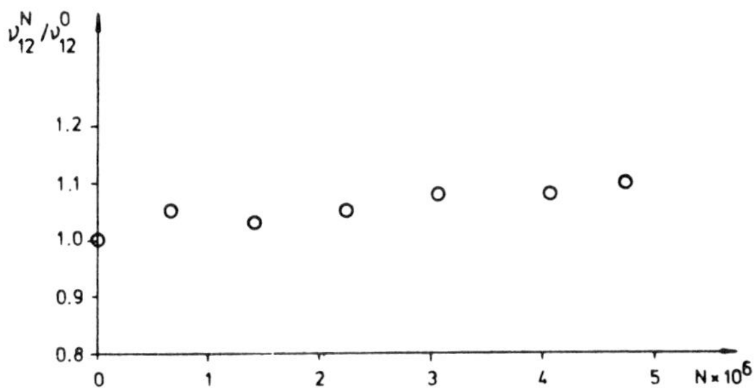

Figure 4. Variation of ν_{12}.

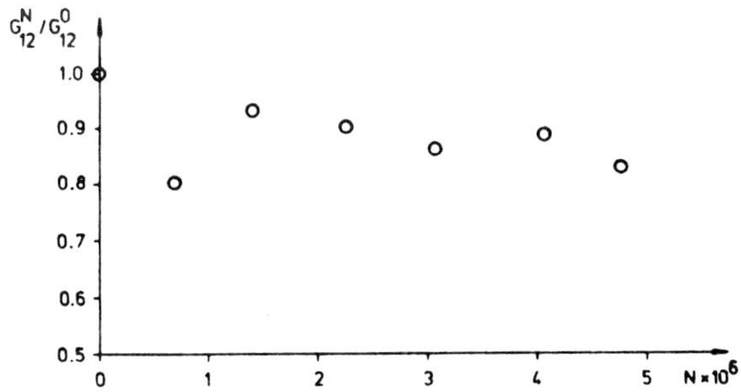

Figure 5. Variation of G_{12}.

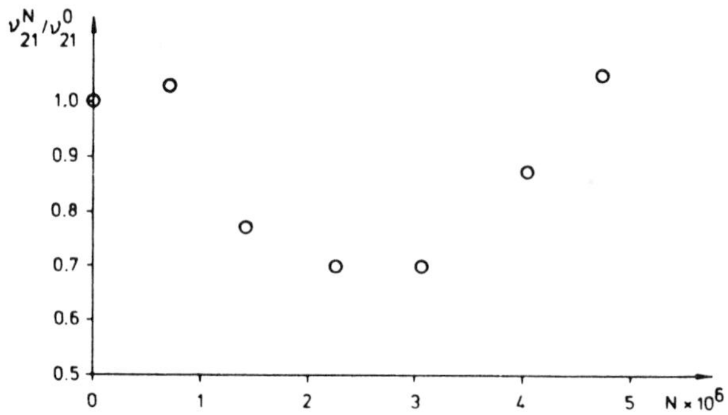

Figure 6. Variation of ν_{21}.

shear modulus G_{12} and the Poisson's ratio ν_{21} show significant changes.

It is believed that the degradation observed in both G_{12} and ν_{21} reflects matrix damage. No splitting parallel to the fibres was observed in the specimen under consideration which indicates absence of an extensive fibre breakage. Previous studies, e.g. [2], have indicated that fatigue damage is dominated by fibre breakage at high stresses while at low stresses matrix microcracking is the major damage mechanism. Due to the low stresses applied to the specimen in this investigation it may be expected that matrix cracking would dominate the fatigue damage. The variation in the shear modulus seems to indicate that the matrix damage might approach a saturation state. Since E_{11} and ν_{12} show small changes, the changes in E_{22} are essentially reflected by the changes in ν_{21}.

From the results described above it is apparent that changes in E_{11} alone do not reflect all damage events in fatigue of composites. Therefore, a damage characterization of composites cannot be based on the measured changes in E_{11} alone. A complete stiffness based characterization would require the totality of changes in all stiffness properties. Clearly, these changes must be related to the damage mechanisms, if possible, in a quantitative way. For instance, the matrix microcrack density may be related to the changes in E_{22} and G_{12}. Fibre debonding may affect E_{11} and interfacial damage might influence changes in ν_{12}.

Extension to Laminates

A lamina is the basic element of laminates. The laminated plate theory provides a basis for relating the elastic properties of the individual laminae to the elastic properties of the laminate. It is to be expected, therefore, that changes induced by fatigue in the elastic properties of individual laminae would result in changes of the elastic properties of the laminate. A stiffness based fatigue damage characterization of laminates would therefore require establishing changes in the stiffness properties of the individual laminae induced by fatigue. The individual laminae however do not behave freely but are constrained by the neighbouring laminae. This constraint effect on fatigue induced stiffness property changes should therefore be investigated next.

The procedure described above for unidirectional composites can be extended to other configurations of laminates. One useful configuration to study the constraint effects would be the cross-ply laminates. This configuration is specially suited to study the matrix damage changes as the 90° plies provide the most direct suppression of matrix cracking.

Conclusions

A procedure to investigate the stiffness property changes induced by fatigue of unidirectional composites has been proposed. The results obtained for a unidirectional glass fibre reinforced polyester clearly demonstrate the necessity for considering all four stiffness properties for a complete characterization of the fatigue induced damage.

Acknowledgements

The author wishes to thank A. Lystrup and F. Jensen for fabrication of the specimen material and H. Lilholt for useful discussions.

References

1. Hahn, H. T. and Kim, R. Y., "Fatigue Behavior of Composite Laminate," *Journal of Composite Materials*, 10, pp. 156–180 (1976).
2. Dharan, C. K. H., "Fatigue Failure Mechanisms in a Unidirectionally Reinforced Composite Material," *Fatigue of Composite Materials*, ASTM STP 569, American Society for Testing and Materials, pp. 171–188 (1975).
3. Reifsnider, K. L., Stinchcomb, W. W. and O'Brien, T. K., "Frequency Effects on a Stiffness-Based Fatigue Failure Criterion in Flawed Composite Specimens," *Fatigue of Filamentary Composite Materials*, ASTM STP 636, American Society for Testing and Materials, pp. 171–184 (1977).
4. Kim, H. C. and Ebert, L. J., "Axial Fatigue Failure Sequence and Mechanisms in Unidirectional Fiberglas Composites," *Journal of Composite Materials*, 12, pp. 139–152 (1978).
5. O'Brien, T. K. and Reifsnider, K. L., "A Complete Mechanical-Property Characterization of a Single Composite Specimen," *Experimental Mechanics*, 20, No. 5, pp. 145 152 (1980).

CHAPTER 6

A Continuum Mechanics Characterization of Damage in Composite Materials

Abstract: The mechanical response of a composite material damaged in fatigue, or in any other loading mode, is characterized by means of a set of vector fields, each representing a damage mode. Constitutive equations are derived for isothermal small-deformation behaviour following attainment of a damage state. For small damage, the elastic constants of a damage state are related to those of the undamaged state. The usefulness of these relations in characterizing the development of damage, for instance during fatigue, is discussed.

FOR MORE THAN a decade, considerable efforts have been made in seeking clarification of the mechanisms involved in the damage of composite materials. In particular, numerous investigations of fatigue damage have been made (Reifsnider 1979, Stinchcomb and Reifsnider 1979, and Hahn 1979). A large body of information has been collected utilizing various nondestructive techniques, e.g. ultrasonic C-scan, simple and stero x-ray radiography, edge replication and thermovibrography, and destructive techniques, e.g. specimen sectioning and de-ply for microscopy observations (Henneke and Reifsnider 1979, Henneke et al. 1979, and Jamison 1982). The picture of damage which has emerged shows a multiplicity of cracking modes and their three-dimensional distribution. In an effort to understand the evolution of damage, a systematization of the damage mechanisms has been proposed, which leads to a consistent pattern of fatigue life diagrams, as discussed in Chapter 2.

In the light of the information available on damage mechanisms, it can hardly be disputed that a single crack characterization of damage would be far removed from reality. For a realistic and rational description of damage, what we need is a model, which is more general than a single crack and which contains essential features of the damage mechanisms which are known from experimental observations.

Reprinted by permission of the Royal Society of London.

In the present chapter, we shall present a continuum mechanics model for damage, which incorporates damage in the material response functions as a set of internal vector field variables. We shall then derive the constitutive equations for isothermal, small-deformation behavior of composite materials in a given damage state. For small damage, the residual elastic properties will be derived and related to the initial elastic properties. The usefulness of these relations to the assessment of damage development will then be discussed.

Damage

We shall define "damage" as a collection of permanent microstructural changes brought about in a material by a set of irreversible physical (or chemical) processes resulting from the application of thermomechanical loadings.

In advanced composites, e.g. graphite-epoxy, the microstructural changes are in the form of cracks of various geometries and orientations, e.g. matrix cracks, fiber breaks, fiber/matrix debonds and interlaminar cracks. The development of the microstructural changes is conveniently observed in fatigue, since the rate of change is small compared to that in, for instance, quasi-static fracture and impact failure.

Based on fatigue damage studies of advanced composite laminates, the development of damage may be described as a two-stage process. The first stage consists essentially of multiple cracking of the matrix, which eventually saturates. The saturation state has been termed a "characteristic damage state," or CDS (Reifsnider and Talug 1978), and the stress fields in the vicinity of cracks in this state have been studied (Talug and Reifsnider 1979). The spacings of cracks have been predicted for bidirectional laminates by Aveston and Kelly (1973, 1980) and for multidirectional laminates by Reifsnider and Highsmith (1981).

The second stage of damage is less well understood and has not been collectively subjected to an analytical treatment. Observations show the appearance of secondary cracks followed by delaminations in the vicinity of previously formed (primary) matrix cracks (Jamison 1982). In contrast to the first stage, some localization, in the sense of damage development at preferred sites, appears to take place. The final fracture seems either to result from severe interactions of cracks leading to loss of material integrity or to be controlled by fibre failures. The fibre-controlled fracture has been the subject of several stochastic studies (see, for instance, Harlow and Phoenix 1978). However, much needs to be done experimentally to clarify the nature of this process.

For the sake of continuum mechanics modeling of the overall material response, let us consider the following two properties of damage in composite materials.

1. Damage occurs in the form of different multiple cracking modes. There is

no isolated single crack that dominates the development of damage. (Free-edge delaminations are not relevant damage modes, as these are boundary layer effects and must therefore be considered separately.)

2. In composite laminates, damage develops along preferred orientations. For instance, matrix cracks in off-axis plies are usually channeled by the fibres in the plies, and interlaminar planar cracks grow along fibres in the neighboring plies.

Taking the first property of damage into consideration, it becomes apparent that a crack-based fracture mechanics type approach to modeling would require the determination of stress fields around complex crack patterns for each damage mode. This may not always be tractable. As an alternative approach, we may consider the effect of a multitude of cracks smeared into a locally homogeneous field and then take the field quantities of this field to characterize the damage state. This approach has been used, for instance, in the context of dislocations for plastic deformation (Kratochvil and Dillon 1969), voids for creep (Leckie 1978) and microcracks for brittle failure (Davison and Stevens 1973, Krajcinovic and Fosenka 1981).

The second property of damage suggests that an adequate characterization of damage would require a continuum vector field. A scalar field would suffice if the orientations of cracks have no effect on the material response. However, in the case of highly anisotropic materials, such as advanced composites, a significant effect of crack orientation on the overall response is to be expected. A second-order or higher order tensor field characterization would become necessary if the effect of crack surface motions on the material response were significant. However, we shall restrict ourselves to small-deformation response and be content with a vector field characterization for damage.

Consider an infinitesimal volume dv surrounding a point P in a damaged solid (Figure 1). Let this volume contain m planar cracks. To each crack, let us assign a vector $\mathbf{p}^{(r)}$, $r = 1,2, \ldots m$, whose magnitude is given by a func-

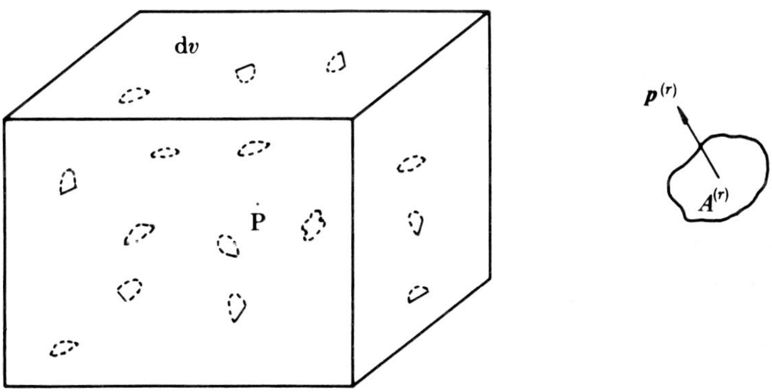

Figure 1. Vector field characterization of damage.

tion of the crack area $A^{(r)}$ and a characteristic crack dimension, $a^{(r)}$, and whose direction is normal to the crack plane. Thus, we have,

$$\mathbf{p}^{(r)} = d^{(r)}\mathbf{n}^{(r)} \tag{2.1}$$

where

$$d^{(r)} = f\{A^{(r)}, a^{(r)}\} \tag{2.2}$$

and $\mathbf{n}^{(r)}$ is a unit vector normal to the rth crack plane.

We shall assume that there exist n crack plane orientations in the volume dv. For each orientation, then, we may define a vector $\mathbf{V}^{(\alpha)}$, $\alpha = 1,2, \ldots n$, given by

$$\mathbf{V}^{(\alpha)} = D^{(\alpha)}\mathbf{n}^{(\alpha)} \tag{2.3}$$

where $\mathbf{n}^{(\alpha)}$ is a unit vector normal to the crack planes of the αth orientation, and $D^{(\alpha)}$ is a suitably defined average over the volume dv of all $d^{(r)}$ corresponding to $\mathbf{n}^{(r)} = \mathbf{n}^{(\alpha)}$.

The function f in Equation (2.2) and the averaging procedure for determining D are kept arbitrary at this point, as the theoretical treatment of the constitutive equations in the following section is not affected by the choice of these. However, for the sake of illustration of the concepts involved, we shall consider a specific example below in relation to the correspondence between the scalar field characterization and the vector field characterization.

SCALAR FIELD CHARACTERIZATION

The most familiar scalar field characterization is due to Kachanov (1958), who proposed taking the crack number density as the scalar value of damage in cracked bodies. We shall see below that a specific choice of the function f and a specific averaging procedure for determining D lead to a relation between the resulting vector field value and the crack number density.

Let us first assume that the scalar field value at a point P in a cracked body can be expressed as a polynomial function of the set of vectors $\mathbf{V}^{(\alpha)}$. Thus, let

$$S = F(\mathbf{V}^{(\alpha)}) \tag{2.4}$$

where S is the scalar field value at the point P and F stands for a polynomial function.

We can express F in terms of the invariants of the vector set $\mathbf{V}^{(\alpha)}$. The invariants are given by (see, for example, Weyl, 1946),

$$V_i^{(\alpha)} V_i^{(\beta)} \text{ and } e_{ijk}V_i^{(\alpha)} V_j^{(\beta)} V_k^{(\gamma)} \tag{2.5}$$

where $\mathbf{V}_i^{(\alpha)}$ are the Cartesian components of $\mathbf{V}^{(\alpha)}$ and e_{ijk} are components of the alternating tensor.

For one orientation of cracks, i.e. $\alpha = 1$, we have, from Equations (2.3) and (2.5),

$$S = F(V_i V_i) = F(D^2). \tag{2.6}$$

Consider now the function f in Equation (2.2) to be given by

$$f = \{c.\ A^{(r)}.\ a^{(r)}\}^{1/2} \tag{2.7}$$

where c is a crack shape parameter defined in the range $0 \leq c \leq 1$. Further, let us use the following relation for determining D.

$$\mathbf{V} \cdot \mathbf{V} = D^2 = \frac{1}{dv} \Sigma \mathbf{p}^{(r)} \cdot \mathbf{p}^{(r)} = \frac{1}{dv} \Sigma \{d^{(r)}\}^2 \tag{2.8}$$

From Equations (2.2), (2.7) and (2.8), we obtain,

$$D^2 = \eta_c < c.A.a > \tag{2.9}$$

where η_c is the crack number density and the symbol $< \ >$ stands for the average of the quantities within it.

For dilute concentration of cracks, $D^2 << 1$, which, from Equation (2.6) yields the approximate relation,

$$S = k\,D^2 \tag{2.10}$$

where k is a constant.

Taking the crack number density as the scalar field value, in accordance with Kachanov, i.e. for $S = \eta$, we have,

$$k = \frac{1}{<cAa>} \tag{2.11}$$

Constitutive Equations

Let us consider a composite material laminate, initially stress-free and undamaged, being subjected to a given loading. Let the loading be interrupted at time t and let the material rest until it attains a natural state, i.e. a state of thermodynamically stable equilibrium. Following this, let mechanical loads be applied to the material, isothermally and gradually, such that the material suffers only small strains. We shall seek a description of material response under this loading, assuming that the material behaves elastically under the loading and that it suffers no further damage during the loading.

Following the approach to thermodynamics of continua proposed by Coleman and Noll (1963) and applied to phenomenological constitutive equations with internal variables by Coleman and Gurtin (1967), we assume that the material response is given by the following functions:

The stress tensor, σ,
the heat flux vector, $\mathbf{q}$,
the specific Helmholz free energy, ψ,
and the specific entropy, η,

and that these functions are determined by the following set of state variables:

The temperature, τ,
the deformation gradient, $\mathbf{F}$,
the temperature gradient, g,
and the damage vectors, $\mathbf{V}^{(\alpha)}$.

The rate of damage vector, $\dot{\mathbf{V}}^{(\alpha)}$ which was also taken as a response function by Coleman and Gurtin is not included in the set of response functions here, as we are not seeking a description of the damage growth but only a description of the elastic response following attainment of a damage state.

We assume, in accordance with the principle of equipresence proposed by Truesdell and Toupin (1960) that each response function is a function of all state variables.

As a consequence of the second law of thermodynamics, in the form of the Clausius-Duhem inequality, which states that the rate of production of entropy is not negative, the response functions σ, ψ and η become independent of the temperature gradient (Coleman and Gurtin 1967). Thus we have

$$\sigma = \sigma(\mathbf{F},\tau,\mathbf{V}^{(\alpha)}) \tag{3.1}$$

$$\psi = \psi(\mathbf{F},\tau,\mathbf{V}^{(\alpha)}) \tag{3.2}$$

$$\eta = \eta(\mathbf{F},\tau,\mathbf{V}^{(\alpha)}) \tag{3.3}$$

and

$$\mathbf{q} = \mathbf{q}(\mathbf{F},\tau,g,\mathbf{V}^{(\alpha)}) \tag{3.4}$$

As a further consequence of the Clausius-Duhem inequality we have,

$$\sigma = \varrho \frac{\partial \psi}{\partial \mathbf{F}}(\mathbf{F},\tau,\mathbf{V}^{(\alpha)}) \tag{3.5}$$

and

$$\eta = -\frac{\partial\psi}{\partial\tau}(\mathbf{F},\tau,\mathbf{V}^{(\alpha)}) \tag{3.6}$$

where ϱ is the mass density.

Restricted to isothermal, small elastic deformations, the response functions are given by

$$\sigma = \sigma(\mathbf{e},\mathbf{V}^{(\alpha)}) \tag{3.7}$$

and

$$\psi = \psi(\mathbf{e},\mathbf{V}^{(\alpha)}) \tag{3.8}$$

together with the relation

$$\sigma = \varrho\frac{\partial\psi}{\partial\mathbf{e}}(\mathbf{e},\mathbf{V}^{(\alpha)}) \tag{3.9}$$

where **e** is the second-order elastic strain tensor.

MATERIAL SYMMETRY RESTRICTIONS

In most applications, composite materials are composed to have certain symmetry properties. The material response functions must comply with the symmetry properties by remaining invariant under coordinate transformations that express the symmetry properties.

Rivlin and his co-workers investigated the restrictions imposed by material symmetry properties on response functions that can be approximated by polynomials (Rivlin and Eriksen 1955; Rivlin 1955; Pipkin and Rivlin 1959). They used the classical invariant theory, according to which any polynomial scalar function of any number of tensors, which is invariant under a given orthogonal group of coordinate transformations, can be expressed as a polynomial in a finite number of polynomial scalar invariants, none of which is expressible as a polynomial in the remaining ones. Such a set of polynomial scalar invariants is called an irreducible integrity basis for the tensors under the given orthogonal group of coordinate transformations. The restrictions imposed by the form-invariance on a response function is thus reduced to determining an irreducible integrity basis for the argument tensors under a group of orthogonal transformations associated with the material symmetry.

Irreducible integrity bases for vectors and tensors of second-order under various symmetry transformations have been determined (Adkins 1960, Smith et al. 1963; Smith and Rivlin, 1964). An account of all integrity bases found until 1964 is given by Wineman and Pipkin (1964). Further results have been published most recently by Smith (1982).

The response function of our interest, ψ, is a function of the second-order symmetric tensor components e_{ij} and vector components $V_k^{(\alpha)}$. We shall consider the restrictions imposed on its form for two commonly encountered composite material symmetries.

TRANSVERSELY ISOTROPIC MATERIALS

The basic element of most composite laminates is a unidirectional ply. A unidirectionally reinforced material may be considered isotropic in a plane normal to the fiber direction. For transversely isotropic symmetry, the integrity basis for a symmetric second-order tensor and n vectors has been determined by Pipkin and Rivlin (1959). Taking the coordinate axis x_1 as the axis of symmetry, the orthogonal group of transformations consists of all rotations about x_1 or reflections in any plane containing x_1. The elements of the integrity basis are the following invariants.

$$\begin{aligned} & e_{11},\ e_{pp},\ e_{pq}e_{pq}, \\ & e_{1p}e_{p1},\ e_{1p}e_{pq}e_{q1}, \\ & V_1^{(\alpha)},\ V_p^{(\alpha)}e_{p1},\ V_p^{(\alpha)}\ e_{pq}e_{q1}, \\ & V_p^{(\alpha)}\ V_p^{(\beta)},\ V_p^{(\alpha)}\ e_{pq}V_q^{(\beta)} \end{aligned} \tag{3.10}$$

where $\alpha,\beta = 1,2$; $p,q = 2,3$ and repeated indices imply summation.

For one vector field, $\mathbf{V}$, in a transversely isotropic material subjected to isothermal, small strains, the Helmholtz free energy can be written as (see Appendix A)

$$\varrho\psi = \varrho\psi^0 + \varrho\psi^1 \tag{3.11}$$

Applying Equation (3.9) to Equation (3.11), we get

$$\sigma_{ij} = \varrho\frac{\partial\psi^0}{\partial e_{ij}} + \varrho\frac{\partial\psi^1}{\partial e_{ij}} \tag{3.12}$$

Using Equations (A.8–A.10) of Appendix A in Equation (3.12), we obtain

$$\sigma_{ij} = (C^0_{ijk\ell} + C^1_{ijk\ell})e_{k\ell} \tag{3.13}$$

which, on using the Voigt notation can be rewritten as

$$\sigma_p = (C^0_{pq} + C^1_{pq})e_q \tag{3.14}$$

where the coefficients C^0_{pq} are given by

$$
\begin{aligned}
C^0_{11} &= 2A_1 \\
C^0_{12} &= C^0_{21} = C^0_{13} = C^0_{31} = A_2 \\
C^0_{14} &= C^0_{41} = C^0_{15} = C^0_{51} = C^0_{16} = C^0_{61} = 0 \\
C^0_{22} &= 2(B_1 + C_1) = C^0_{33} \\
C^0_{23} &= C^0_{32} = 2B_1 \\
C^0_{24} &= C^0_{42} = C^0_{25} = C^0_{52} = C^0_{26} = C^0_{62} = 0 \\
C^0_{34} &= C^0_{43} = C^0_{35} = C^0_{53} = C^0_{36} = C^0_{63} = 0 \\
C^0_{44} &= 2C_1 \\
C^0_{45} &= C^0_{54} = C^0_{46} = C^0_{64} = 0 \\
C^0_{55} &= C^0_{66} = D_1,
\end{aligned}
\tag{3.15}
$$

and the coefficients C^1_{pq} are given by

$$
\begin{aligned}
C^1_{11} &= 2(A_3V_1^2 + A_5V_2^2 + A_5V_3^2) \\
C^1_{12} &= C^1_{21} = A_6V_2^2 + A_7V_1^2 + A_8V_2^2 + A_8V_3^2 \\
C^1_{13} &= C^1_{31} = A_6V_3^2 + A_7V_1^2 + A_8V_2^2 + A_8V_3^2 \\
C^1_{14} &= C^1_{41} = A_6V_2V_3 \\
C^1_{15} &= C^1_{51} = C^1_{16} = C^1_{61} = 0 \\
C^1_{22} &= 2(B_2V_1^2 + B_4V_2^2 + B_4V_3^2 + B_5V_2^2 + C_2V_1^2 + C_3V_2^2 + C_3V_3^2) \\
C^1_{23} &= C^1_{32} = 2B_2V_1^2 + 2B_4V_2^2 + 2B_4V_3^2 + B_5V_3^2 + B_5V_2^2 \\
C^1_{24} &= C^1_{42} = B_5V_2V_3 \\
C^1_{25} &= C^1_{52} = 0 \\
C^1_{26} &= C^1_{62} = 1/2\ E_1V_1V_2 \\
C^1_{33} &= 2(B_2V_1^2 + B_4V_2^2 + B_4V_3^2 + B_5V_3^2 + C_2V_1^2 + C_3V_2^2 + C_3V_3^2) \\
C^1_{34} &= C^1_{43} = B_5V_2V_3 \\
C^1_{35} &= C^1_{53} = 1/2\ E_1V_1V_3 \\
C^1_{36} &= C^1_{63} = 0 \\
C^1_{44} &= C_2V_1^2 + C_3(V_2^2 + V_3^2) \\
C^1_{45} &= C^1{}_{54} = 1/4\ E_1V_2V_1 \\
C^1_{46} &= C^1_{64} = 1/4\ E_1V_1V_3 \\
C^1_{55} &= 1/2\ (D_2V_1^2 + D_3V_2^2 + D_3V_3^2 + F_1V_3^2) \\
C^1_{56} &= C^1_{65} = 1/2\ F_1V_2V_3 \\
C^1_{66} &= 1/2\ (D_2V_1^2 + D_3V_2^2 + D_3V_3^2 + F_1V_2^2)
\end{aligned}
\tag{3.16}
$$

From Equations (3.15) it can be seen that the coefficient matrix C^0_{pq} is the familiar elastic coefficient matrix for the generalized Hooke's law for transversely isotropic materials. The coefficient matrix C^1_{pq} is symmetric and has the following form.

$$
C^1_{pq} = \begin{bmatrix}
C^1_{11} & C^1_{12} & C^1_{13} & C^1_{14} & 0 & 0 \\
 & C^1_{22} & C^1_{23} & C^1_{24} & 0 & C^1_{26} \\
 & & C^1_{33} & C^1_{34} & C^1_{35} & 0 \\
 & & & C^1_{44} & C^1_{45} & C^1_{46} \\
 & & & & C^1_{55} & C^1_{56} \\
 & & & & & C^1_{66}
\end{bmatrix}
\tag{3.17}
$$

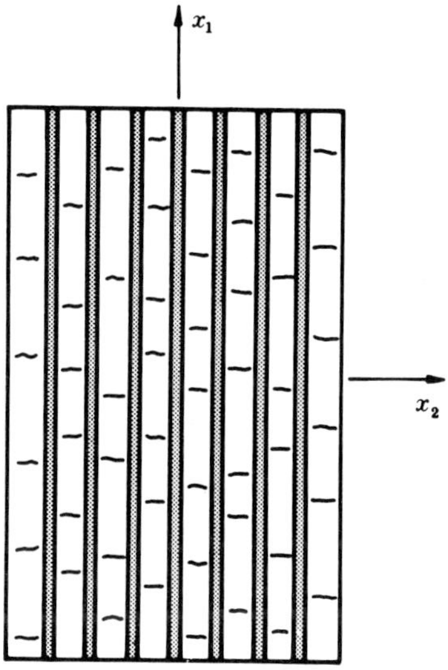

Figure 2. Matrix cracking normal to fibres in a unidirectional composite.

Furthermore, from Equation (3.16) we have,

$$C^1_{24} = C^1_{34} \quad \text{and} \quad C^1_{35} = C^1_{46} \tag{3.18}$$

In unidirectional composites, damage occurs as cracks normal to and parallel to the fibers. Let us first consider cracks normal to fibers. The damage vector is given by (see Figure 2).

$$\mathbf{V} = D_N(1,0,0) \tag{3.19}$$

The coefficient matrix C^1_{pq} for this damage takes the form

$$C^1_{pq} = \begin{bmatrix} C^1_{11} & C^1_{12} & C^1_{13} & 0 & 0 & 0 \\ & C^1_{22} & C^1_{23} & 0 & 0 & 0 \\ & & C^1_{33} & 0 & 0 & 0 \\ & & & C^1_{44} & 0 & 0 \\ & & & & C^1_{55} & 0 \\ & & & & & C^1_{66} \end{bmatrix} \tag{3.20}$$

and

$$C^1_{12} = C^1_{13}, \quad C^1_{22} = C^1_{33}, \quad \text{and} \quad C^1_{55} = C^1_{66} \tag{3.21}$$

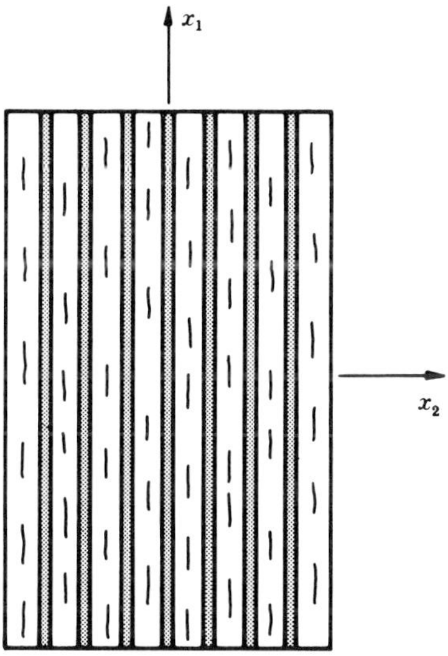

Figure 3. Matrix cracking parallel to fibres in a unidirectional composite.

where

$$\begin{aligned} C^1_{11} &= 2A_3D_N^2 \\ C^1_{12} &= A_7D_N^2 \\ C^1_{22} &= (2B_2 + C_2)D_N^2 \\ C^1_{23} &= 2B_2D_N^2 \\ C^1_{44} &= 1/2\ C_2D_N^2 \\ \text{and} \qquad C^1_{55} &= 1/4\ D_2D_N^2 \end{aligned} \tag{3.22}$$

From Equations (3.20–3.22) it can be seen that the coefficient matrix C^1_{pq} for this damage has the same form as the coefficient matrix C^0_{pq}. Thus the transverse isotropic symmetry in the elastic behavior is preserved under this damage.

Let us next consider damage of the type shown in Figure 3, which is given by

$$\mathbf{V} = D_p(0,1,0) \tag{3.23}$$

The coefficient matrix C^1_{pq} for this damage has the following form

$$C^1_{pq} = \begin{bmatrix} C^1_{11} & C^1_{12} & C^1_{13} & 0 & 0 & 0 \\ & C^1_{22} & C^1_{23} & 0 & 0 & 0 \\ & & C^1_{33} & 0 & 0 & 0 \\ & & & C^1_{44} & 0 & 0 \\ & & & & C^1_{55} & 0 \\ & & & & & C^1_{66} \end{bmatrix} \tag{3.24}$$

where

$$
\begin{aligned}
C^1_{11} &= 2A_5D^2_P \\
C^1_{12} &= (A_6 + A_8)D^2_P \\
C^1_{13} &= A_8D^2_P \\
C^1_{22} &= (2B_4 + 2B_5 + 2C_3)D^2_P \\
C^1_{23} &= (2B_4 + B_5)D^2_P \\
C^1_{33} &= 2(B_4 + C_3)D^2_P \\
C^1_{55} &= 1/2\ D_3D^2_P
\end{aligned}
$$

and

$$
C^1_{66} = 1/2\ (D_3 + F_1)D^2_P \tag{3.25}
$$

It can be seen from Equation (3.24) that the coefficient matrix C^1_{pq} under this damage has the form of an elastic coefficient matrix for orthotropic materials.

ORTHOTROPIC MATERIALS

Most composite laminates for practical applications have orthotropic symmetry. The integrity bases for a symmetric second-order tensor and a vector have been determined by Smith et al. (1963) for various crystal classes. The crystal class of interest here is the rhombic-dipyramidal system defined by the group of transformations: $(I,C,R_1,R_2,R_3,D_1,D_2,D_3)$, where

$$
\begin{aligned}
I &= (1,1,1), & C &= (-1,-1,-1) & & \\
R_1 &= (-1,1,1), & R_2 &= (1,-1,1), & R_3 &= (1,1,-1) \\
D_1 &= (1,-1,-1), & D_2 &= (-1,1,-1) & \text{and } D_3 &= (-1,-1,1)
\end{aligned}
\tag{3.26}
$$

The transformation I is the identity transformation, C is the central inversion, R_1, R_2 and R_3 are reflections in planes normal to x_1-, x_2- and x_3-axes, respectively, and D_1, D_2 and D_3 are rotations through 180° of the coordinate system about the x_1-, x_2- and x_3-axes, respectively.

The invariants for this system are given by (Smith *et al.* 1963),

$$
\begin{aligned}
I_1 &= e_{11} \\
I_2 &= e_{22} \\
I_3 &= e_{33} \\
I_4 &= e^2_{23} \\
I_5 &= e^2_{13} \\
I_6 &= e^2_{12} \\
I_7 &= e_{23}e_{13}e_{12} \\
I_8 &= V^2_1 \\
I_9 &= V^2_2 \\
I_{10} &= V^2_3 \\
I_{11} &= V_2V_3e_{23}
\end{aligned}
$$

$$
\begin{aligned}
I_{12} &= V_3 V_1 e_{31} \\
I_{13} &= V_1 V_2 e_{12} \\
I_{14} &= V_2 V_3 e_{13} e_{12} \\
I_{15} &= V_3 V_1 e_{12} e_{23} \\
I_{16} &= V_1 V_2 e_{23} e_{13}
\end{aligned} \qquad (3.27)
$$

The coefficient matrices C^0_{pq} and C^1_{pq} are in this case given by the following. (See Appendix B for the expressions for the Helmholz free energy.)

$$
C^0_{pq} = \begin{bmatrix}
2a_1 & a_2 & a_3 & 0 & 0 & 0 \\
 & 2b_1 & b_2 & 0 & 0 & 0 \\
 & & 2c_1 & 0 & 0 & 0 \\
 & & & \frac{d_1}{2} & 0 & 0 \\
 & & & & \frac{e_1}{2} & 0 \\
 & & & & & \frac{f_1}{2}
\end{bmatrix} \qquad (3.28)
$$

and

$$
\begin{aligned}
C^1_{11} &= 2a_4\, V_1^2 + 2a_5\, V_2^2 + 2a_6\, V_3^2 \\
C^1_{12} &= C^1_{21} = a_{10}\, V_1^2 + a_{11}\, V_2^2 + a_{12}\, V_3^2 \\
C^1_{13} &= C^1_{31} = a_{13}\, V_1^2 + a_{14}\, V_2^2 + a_{15}\, V_3^2 \\
C^1_{14} &= C^1_{41} = 1/2\; a_7\, V_2\, V_3 \\
C^1_{15} &= C^1_{51} = 1/2\; a_8\, V_1\, V_3 \\
C^1_{16} &= C^1_{61} = 1/2\; a_9\, V_1\, V_2 \\
C^1_{22} &= 2b_3\, V_1^2 + 2b_4\, V_2^2 + 2b_5\, V_3^2 \\
C^1_{23} &= C^1_{32} = b_9\, V_1^2 + b_{10}\, V_2^2 + b_{11}\, V_3^2 \\
C^1_{24} &= C^1_{42} = 1/2\; b_6\, V_2\, V_3 \\
C^1_{25} &= C^1_{52} = 1/2\; b_7\, V_1\, V_3 \\
C^1_{26} &= C^1_{62} = 1/2\; b_8\, V_1\, V_2 \\
C^1_{33} &= 2c_2\, V_1^2 + 2c_3\, V_2^2 + 2c_4\, V_3^2 \\
C^1_{34} &= C^1_{43} = 1/2\; c_5\, V_2\, V_3 \\
C^1_{35} &= C^1_{53} = 1/2\; c_6\, V_1\, V_3 \\
C^1_{36} &= C^1_{63} = 1/2\; c_7\, V_1\, V_2 \\
C^1_{44} &= 1/2\; d_2\, V_1^2 + 1/2\; d_3\, V_2^2 + 1/2\; d_4\, V_3^2 \\
C^1_{45} &= C^1_{54} = 1/4\; g_3\, V_1\, V_2 \\
C^1_{46} &= C^1_{64} = 1/4\; g_2\, V_1\, V_3
\end{aligned}
$$

$$
\begin{aligned}
C^1_{55} &= 1/2\ e_2\ V_1^2 + 1/2\ e_3\ V_2^2 + 1/2\ e_4\ V_3^2 \\
C^1_{56} &= C^1_{65} = 1/4\ g_1\ V_2\ V_3 \\
C^1_{66} &= 1/2\ f_2\ V_1^2 + 1/2\ f_3\ V_2^2 + 1/2\ f_4\ V_3^2
\end{aligned}
\tag{3.29}
$$

The coefficient matrix C^0_{pq} corresponds to the elastic coefficient matrix of the generalized Hooke's law for orthotropic materials. The coefficient matrix C^1_{pq} is symmetric and none of its elements are zero.

For the special case of damage given by

$$
\mathbf{V} = D_I(1,0,0) \tag{3.30}
$$

the coefficient matrix C^1_{pq} is given by

$$
C^1_{pq} = \begin{bmatrix}
2a_4\ D_I^2 & a_{10}\ D_I^2 & a_{13}\ D_I^2 & 0 & 0 & 0 \\
 & 2b_3\ D_I^2 & b_9\ D_I^2 & 0 & 0 & 0 \\
 & & 2c_2\ D_I^2 & 0 & 0 & 0 \\
 & & & 1/2\ d_2\ D_I^2 & 0 & 0 \\
 & & & & 1/2\ e_2\ D_I^2 & 0 \\
 & & & & & 1/2\ f_2\ D_I^2
\end{bmatrix}
\tag{3.31}
$$

The coefficient matrix C^1_{pq} in (3.31) has the same form as the undamaged elastic coefficient matrix C^0_{pq} in (3.28). Thus, the elastic behaviour under the damage given by (3.30) retains orthotropic symmetry.

It can easily be shown that any damage field that is parallel to any of the three orthogonal axes of material symmetry preserves the orthotropic symmetry in the elastic coefficients.

Let us now consider a damage vector normal to an off-axis ply in an orthotropic laminate. Typically, such a damage would represent matrix cracking in the off-axis ply (see Figure 4).

The damage vector is now given by

$$
\mathbf{V} = D_\theta(\sin\theta, -\cos\theta, 0) \tag{3.32}
$$

The coefficient matrix C^1_{pq} now takes the following form:

$$
C^1_{pq} = \begin{bmatrix}
C^1_{11} & C^1_{12} & C^1_{13} & 0 & 0 & C^1_{16} \\
 & C^1_{22} & C^1_{23} & 0 & 0 & C^1_{26} \\
 & & C^1_{33} & 0 & 0 & C^1_{36} \\
 & & & C^1_{44} & C^1_{45} & 0 \\
 & & & & C^1_{55} & 0 \\
 & & & & & C^1_{66}
\end{bmatrix}
\tag{3.33}
$$

Figure 4. Matrix cracking in an off-axis ply of an orthotropic laminate.

where

$$
\begin{aligned}
C^1_{11} &= 2a_4\, D^2_\theta \sin^2\theta + 2a_5\, D^2_\theta \cos^2\theta \\
C^1_{12} &= a_{10}\, D^2_\theta \sin^2\theta + a_{11}\, D^2_\theta \cos^2\theta \\
C^1_{13} &= a_{13}\, D^2_\theta \sin^2\theta + a_{14}\, D^2_\theta \cos^2\theta \\
C^1_{16} &= -\ 1/2\ a_9\, D^2_\theta \sin\theta \cos\theta \\
C^1_{22} &= 2b_3\, D^2_\theta \sin^2\theta + 2b_4\, D^2_\theta \cos^2\theta \\
C^1_{23} &= b_9\, D^2_\theta \sin^2\theta + b_{10}\, D^2_\theta \cos^2\theta \\
C^1_{26} &= -\ 1/2\ b_8\, D^2_\theta \sin\theta \cos\theta \\
C^1_{33} &= 2c_2\, D^2_\theta \sin^2\theta + 2c_3\, D^2_\theta \cos^2\theta \\
C^1_{36} &= -\ 1/2\ c_7\, D^2_\theta \sin\theta \cos\theta \\
C^1_{44} &= 1/2\ d_2\, D^2_\theta \sin^2\theta + 1/2\ d_3\, D^2_\theta \cos^2\theta \\
C^1_{45} &= -\ 1/4\ g_3\, D^2_\theta \sin\theta \cos\theta \\
C^1_{55} &= 1/2\ e_2\, D^2_\theta \sin^2\theta + 1/2\ e_3\, D^2_\theta \cos^2\theta \\
C^1_{66} &= 1/2\ f_2\, D^2_\theta \sin^2\theta + 1/2\ f_3\, D^2_\theta \cos^2\theta
\end{aligned}
\qquad (3.34)
$$

Equation (3.33) shows that the elastic coefficients for this damage acquire monoclinic symmetry with the coordinate plane x_1x_2 as the plane of symmetry. It can easily be shown that a damage vector with $V_i \neq 0$, $V_j \neq 0$, $V_k = 0$ in an orthotropic material will lead to monoclinic symmetry in the elastic coefficients with respect to the x_ix_j plane.

MULTIPLE VECTOR FIELDS

The nature of damage in composite materials is such that different damage modes may develop simultaneously, at different growth rates. At a given time, then, a description of the damage state may require more than a single vector field.

The integrity basis for a second-order symmetric tensor with components e_{ij} and n vectors $\mathbf{V}^{(\alpha)}$, $\alpha = 1,2 \ldots, n$, with components $\mathbf{V}_k^{(\alpha)}$ will, in general, have elements that are functions of the following type.

$$
\begin{array}{ll}
f_1(e_{ij}) & \\
f_2(e_{ij}, V_k^{(\alpha)}) & \\
f_3(V_k^{(\alpha)}) & \\
f_4(e_{ij}, V_k^{(\alpha)}, V_\ell^{(\beta)}) & , \quad \alpha \neq \beta \\
\text{and} \quad f_5(V_i^{(\alpha)}, V_j^{(\beta)}) & , \quad \alpha \neq \beta
\end{array}
\tag{3.35}
$$

If we assume that, in determining the overall response of a material, any two damage vector fields do not interact, then, in forming a response function, we may neglect the functions of the type f_4 and f_5. The response function $\varrho\psi$ may now be written as the following polynomial.

$$\varrho\psi = F_1(e_{ij}) + F_2^{(\alpha)}(e_{ij}, V_k^{(\alpha)}) + F_3^{(\alpha)}(V_i^{(\alpha)}) \tag{3.36}$$

Imposing now the conditions that for an unstrained and undamaged material $\varrho\psi$ vanishes and that an unstrained material of any damage state is stress free, we have, retaining up to second-order terms in e_{ij} and V_k,

$$\varrho\psi = \varrho\psi^0 + \varrho\psi^{(\alpha)} + P^{(\alpha)} \tag{3.37}$$

where $\varrho\psi^0$ is the elastic strain energy of the undamaged material, $\varrho\psi^{(\alpha)}$ are functions of products of e_{ij} and $V_k^{(\alpha)}$, and $P^{(\alpha)}$ are quadratic functions of $V_i^{(\alpha)}$.

Applying the stress formula

$$\sigma_{ij} = \varrho \frac{\partial\psi}{\partial e_{ij}} \tag{3.38}$$

we obtain

$$\sigma_{ij} = (C^0_{ijk\ell} + C^{(\alpha)}_{ijk\ell})e_{k\ell} \tag{3.39}$$

or, in the Voigt notation

$$\sigma_p = (C^0_{pq} + C^{(\alpha)}_{pq})e_q \tag{3.40}$$

C^0_{pq} are the elastic coefficients of an undamaged material and $C^{(\alpha)}_{pq}$ are functions of the damage vector components $V^{(\alpha)}_k$.

TWO DAMAGE FIELDS IN AN ORTHOTROPIC MATERIAL

From structural considerations, composite laminates are often constructed to have symmetrically placed off-axis plies. Examples of such laminates are angle-ply laminates and quasi-isotropic laminates. In this type of laminate, the damage vector fields describing matrix cracking will be symmetrically oriented.

As an example, let us consider two damage vectors given by

$$\mathbf{V}^{(1)} = D_1(\sin\theta, -\cos\theta, 0) \tag{3.41}$$

and

$$\mathbf{V}^{(2)} = D_2(\sin\theta, \cos\theta, 0) \tag{3.42}$$

representing the damage shown in Figure 5.

The coefficient matrix $C^{(\alpha)}_{pq}$ in this case is given by

$$\begin{aligned}
C^{(\alpha)}_{11} &= 2a_4\, D^2_\alpha \sin^2\theta + 2a_5\, D^2_\alpha \cos^2\theta \\
C^{(\alpha)}_{12} &= a_{10}\, D^2_\alpha \sin^2\theta + a_{11}\, D^2_\alpha \cos^2\theta \\
C^{(\alpha)}_{13} &= a_{13}\, D^2_\alpha \sin^2\theta + a_{14}\, D^2_\alpha \cos^2\theta \\
C^{(\alpha)}_{14} &= C^{(\alpha)}_{15} = 0 \\
C^{(\alpha)}_{16} &= (-1)^\alpha/2\; a_9\, D^2_\alpha \sin\theta\cos\theta \\
C^{(\alpha)}_{22} &= 2b_3\, D^2_\alpha \sin^2\theta + 2b_4\, D^2_\alpha \cos^2\theta \\
C^{(\alpha)}_{23} &= b_9\, D^2_\alpha \sin^2\theta + b_{10}\, D^2_\alpha \cos^2\theta \\
C^{(\alpha)}_{24} &= C^{(\alpha)}_{25} = 0 \\
C^{(\alpha)}_{26} &= (-1)^\alpha/2\; b_8\, D^2_\alpha \sin\theta\cos\theta \\
C^{(\alpha)}_{33} &= 2c_2\, D^2_\alpha \sin^2\theta + 2c_3\, D^2_\alpha \cos^2\theta \\
C^{(\alpha)}_{34} &= C^{(\alpha)}_{35} = 0 \\
C^{(\alpha)}_{36} &= (-1)^\alpha/2\; c_7\, D^2_\alpha \sin\theta\cos\theta \\
C^{(\alpha)}_{44} &= 1/2\; d_2\, D^2_\alpha \sin^2\theta + 1/2\; d_3\, D^2_\alpha \cos^2\theta \\
C^{(\alpha)}_{45} &= (-1)^\alpha/4\; g_3\, D^2_\alpha \sin\theta\cos\theta \\
C^{(\alpha)}_{46} &= 0 \\
C^{(\alpha)}_{55} &= 1/2\; e_2\, D^2_\alpha \sin^2\theta + 1/2\; e_3\, D^2_\alpha \cos^2\theta \\
C^{(\alpha)}_{56} &= 0 \\
C^{(\alpha)}_{66} &= 1/2\; f_2\, D^2_\alpha \sin^2\theta + 1/2\; f_3\, D^2_\alpha \cos^2\theta
\end{aligned} \tag{3.43}$$

where $\alpha = 1, 2$.

The coefficient matrix $C^{(\alpha)}_{pq}$ thus gives the elastic coefficients monoclinic symmetry about the plane x_1x_2, the plane in which the two damage vectors lie.

For $D_1 = D_2$, we obtain, from (3.43),

$$C_{16}^{(\alpha)} = C_{26}^{(\alpha)} = C_{36}^{(\alpha)} = C_{45}^{(\alpha)} = 0 \tag{3.44}$$

This leads to the property that any two symmetrically placed damage fields of equal intensity in an orthotropic material retain the orthotropic symmetry of the elastic coefficients.

For the special case of $\theta = 45°$, we have, for each vector,

$$V_1^2 = V_2^2 \tag{3.45}$$

i.e.

$$I_3 = I_9 \tag{3.46}$$

as given by Equations (3.27).

Observing Equations (B.1) of Appendix B we see that we then must have,

$$\begin{array}{lll} a_4 = a_5, & a_{10} = a_{11}, & a_{13} = a_{14} \\ b_3 = b_4, & b_9 = b_{10}, & c_2 = c_3 \\ d_2 = d_3, & e_2 = e_3, \quad \text{and} & f_2 = f_3 \end{array} \tag{3.47}$$

Figure 5. Matrix cracking in two symmetrically placed off-axis plies.

Thus for damage oriented at ± 45° to an axis of symmetry in an orthotropic material, we have, from Equation (3.43),

$$
\begin{aligned}
C_{11}^{(\alpha)} &= 2a_4\, D_\alpha^2 \\
C_{12}^{(\alpha)} &= a_{10}\, D_\alpha^2 \\
C_{13}^{(\alpha)} &= a_{13}\, D_\alpha^2 \\
C_{14}^{(\alpha)} &= C_{15}^{(\alpha)} = 0 \\
C_{16}^{(\alpha)} &= (-1)^\alpha/4\; a_9\, D_\alpha^2 \\
C_{22}^{(\alpha)} &= 2b_3\, D_\alpha^2 \\
C_{23}^{(\alpha)} &= b_9\, D_\alpha^2 \\
C_{24}^{(\alpha)} &= C_{25}^{(\alpha)} = 0 \\
C_{26}^{(\alpha)} &= (-1)^\alpha/4\; b_8\, D_\alpha^2 \\
C_{33}^{(\alpha)} &= 2c_2\, D_\alpha^2 \\
C_{34}^{(\alpha)} &= C_{35}^{(\alpha)} = 0 \\
C_{36}^{(\alpha)} &= (-1)^\alpha/4\; c_7\, D_\alpha^2 \\
C_{44}^{(\alpha)} &= 1/2\; d_2\, D_\alpha^2 \\
C_{45}^{(\alpha)} &= (-1)^\alpha/8\; g_3\, D_\alpha^2 \\
C_{46}^{(\alpha)} &= 0 \\
C_{55}^{(\alpha)} &= 1/2\; e_2\, D_\alpha^2 \\
C_{56}^{(\alpha)} &= 0 \\
C_{66}^{(\alpha)} &= 1/2\; f_2\, D_\alpha^2
\end{aligned} \qquad (3.48)
$$

For the damage of equal intensity in the two plies, i.e. for

$$D_1 = D_2 = D \qquad (3.49)$$

we have, the damage coefficient matrix C_{pq}^D, defined as

$$C_{pq}^D = C_{pq}^{(1)} + C_{pq}^{(2)} \qquad (3.50)$$

given by

$$
C_{pq}^D = D^2 \begin{bmatrix} 4a_4 & 2a_{10} & 2a_{13} & 0 & 0 & 0 \\ & 4b_3 & 2b_9 & 0 & 0 & 0 \\ & & 4c_2 & 0 & 0 & 0 \\ & & & d_2 & 0 & 0 \\ & & & & e_2 & 0 \\ & & & & & f_2 \end{bmatrix} \qquad (3.51)
$$

Residual Elastic Properties

In the constitutive equations formulated above to characterize an isothermal, small -deformation response of composite materials subjected to damage, we have seen that the coefficient matrix C^0_{pq} determines the elastic response of the virgin material, while the coefficient matrix $C^{(\alpha)}_{pq}$ characterizes the change in the elastic response of a damaged material whose damage state is described by a vector field $\mathbf{V}^{(\alpha)}$. The coefficient matrix C_{pq}, given by $C_{pq} = C^0_{pq} + C^{(\alpha)}_{pq}$, characterizes the current elastic response of a damaged composite.

From the coefficient matrix C_{pq}, we can obtain values of the measurable residual elastic constants that can provide a convenient means for monitoring the state of damage in composite materials. In the case of fatigue damage, attempts have been made to measure the residual elastic constants and relate these to the underlying damage mechanisms (O'Brien and Reifsnider 1977, 1981; O'Brien 1978; Talreja 1980). In the following, we shall derive expressions relating the residual elastic constants to the initial elastic constants for small amounts of damage.

Let us consider an orthotropic composite subjected to damage that preserves the orthotropic symmetry of the elastic coefficients. As we have seen above, any damage field parallel to an axis of symmetry, or any two damage fields of equal strength symmetrically oriented about an axis of symmetry will retain the orthotropic symmetry in elastic coefficients. The constitutive equations for in-plane behavior can now be written as follows.

$$\begin{Bmatrix} \sigma_1 \\ \sigma_2 \\ \sigma_6 \end{Bmatrix} = \begin{bmatrix} C_{11} & C_{12} & 0 \\ C_{21} & C_{22} & 0 \\ 0 & 0 & C_{66} \end{bmatrix} \begin{Bmatrix} e_1 \\ e_2 \\ e_6 \end{Bmatrix} \tag{4.1}$$

where $C_{pq} = C^0_{pq} + C^{(\alpha)}_{pq}$.

Inverting (4.1) we have,

$$\begin{Bmatrix} e_1 \\ e_2 \\ e_6 \end{Bmatrix} = \begin{bmatrix} S_{11} & S_{12} & 0 \\ S_{21} & S_{22} & 0 \\ 0 & 0 & S_{66} \end{bmatrix} \begin{Bmatrix} \sigma_1 \\ \sigma_2 \\ \sigma_6 \end{Bmatrix} \tag{4.2}$$

where

$$\begin{aligned} S_{11} &= \frac{C_{22}}{C_{11}C_{22} - C_{12}^2}, \\ S_{12} = S_{21} &= \frac{-C_{12}}{C_{11}C_{22} - C_{12}^2}, \\ S_{22} &= \frac{C_{11}}{C_{11}C_{22} - C_{12}^2}, \end{aligned} \tag{4.3}$$

and

$$S_{66} = \frac{1}{C_{66}}$$

The elastic constants defined by the Young's moduli E_1 and E_2, the shear modulus G_{12} and the two Poisson's ratios, ν_{12} and ν_{21} are given by

$$E_1 = \frac{1}{S_{11}} = \frac{C_{11}C_{22} - C_{12}^2}{C_{22}}$$

$$E_2 = \frac{1}{S_{22}} = \frac{C_{11}C_{22} - C_{12}^2}{C_{11}}$$

$$G_{12} = \frac{1}{S_{66}} = C_{66}$$

$$\nu_{12} = -\frac{S_{12}}{S_{11}} = \frac{C_{12}}{C_{22}}$$

and

$$\nu_{21} = -\frac{S_{12}}{S_{22}} = \frac{C_{12}}{C_{11}} \tag{4.4}$$

Let us for simplicity consider a single damage field given by

$$\mathbf{V}^{(1)} = D_I(1,0,0) \tag{4.5}$$

The coefficients $C_{pq}^{(1)}$ are given by Equation (3.31). Substituting these in the first of Equations (4.4) and using the relation $C_{pq} = C_{pq}^0 + C_{pq}^{(1)}$, we have,

$$E_1^{(1)} = \frac{(C_{11}^0 + 2a_4 D_I^2)(C_{22}^0 + 2b_3 D_I^2) - (C_{12}^0 + a_{10} D_I^2)^2}{C_{22}^0 + 2b_3 D_I^2} \tag{4.6}$$

Assuming $D_I << 1$ and retaining up to second-order terms in D_I, we obtain,

$$E_1^{(1)} - \frac{C_{11}^0 C_{22}^0 - (C_{12}^0)^2 + D_I^2(2a_4 C_{22}^0 + 2b_3 C_{11}^0 - 2a_{10} C_{12}^0)}{C_{22}^0 + 2b_3 D_I^2}$$

Expanding the denominator in the binomial series and once again retaining up to the second-order terms, we get,

$$E_1^{(1)} = E_1^0 + D_I^2\left[2a_4 + 2b_3\left(\frac{C_{12}^0}{C_{22}^0}\right)^2 - 2a_{10}\frac{C_{12}^0}{C_{22}^0}\right] \tag{4.8}$$

On using the relation $C_{12}^0/C_{22}^0 = \nu_{12}^0$, we have,

$$E_1^{(1)} = E_1^0 + D_I^2[2a_4 + 2b_3(\nu_{12}^0)^2 - 2a_{10}\nu_{12}^0] \tag{4.9}$$

Similarly, the relations for the remaining elastic constants are obtained as follows:

$$E_2^{(1)} = E_2^0 + D_I^2[2b_3 + 2a_4(\nu_{21}^0)^2 - 2a_{10}\nu_{21}^0] \tag{4.10}$$

$$\nu_{12}^{(1)} = \nu_{12}^0 + D_I^2\left(\frac{1 - \nu_{12}^0\nu_{21}^0}{E_2^0}\right)(a_{10} - 2b_3\nu_{12}^0) \tag{4.11}$$

$$\nu_{21}^{(1)} = \nu_{21}^0 + D_I^2\left(\frac{1 - \nu_{12}^0\nu_{21}^0}{E_1^0}\right)(a_{10} - 2a_4\nu_{21}^0) \tag{4.12}$$

and

$$G_{12}^{(1)} = G_{12}^0 + 1/2\, f_2\, D_I^2 \tag{4.13}$$

For a damage field given by

$$\mathbf{V}^{(2)} = D_{II}(0,1,0) \tag{4.14}$$

the elastic constant relations become as follows:

$$E_1^{(2)} = E_1^0 + D_{II}^2(2a_5 + 2b_4(\nu_{12}^0)^2 - 2a_{11}\nu_{12}^0) \tag{4.15}$$

$$E_2^{(2)} = E_2^0 + D_{II}^2(2b_4 + 2a_5(\nu_{21}^0)^2 - 2a_{11}\nu_{21}^0) \tag{4.16}$$

$$\nu_{12}^{(2)} = \nu_{12}^0 + D_{II}^2\left(\frac{1 - \nu_{12}^0\nu_{21}^0}{E_2^0}\right)(2a_{11} - 2b_4\nu_{12}^0) \tag{4.17}$$

$$\nu_{21}^{(2)} = \nu_{21}^0 + D_{II}^2\left(\frac{1 - \nu_{12}^0\nu_{21}^0}{E_1^0}\right)(2a_{11} - 2a_5\nu_{21}^0) \tag{4.18}$$

and

$$G_{12}^{(2)} = G_{12}^0 + 1/2\, f_3\, D_{II}^2 \tag{4.19}$$

It should be noted here that Equations (4.9–4.13) and (4.15–4.19) show that all elastic constants change with damage for each damage field considered. To determine the elastic constants in each case, four unknown coefficients are required. Conversely, for each damage state, measurements of the four independent elastic constants will determine the unknown coefficients. Once these coefficients are determined, however, the change in the elastic constants can be related to the magnitude of the damage vector in each case. This provides a basis for monitoring damage through elastic coefficient changes.

It is interesting here to compare the expressions (4.15–4.19) with those derived recently by Gottesman, Hashin and Brull (1980). They have used the stress field solutions for a single crack parallel to a symmetry axis in an orthotropic solid derived by Sih, Paris and Irwin (1965) to determine the effective moduli of a unidirectional composite having a number of cracks parallel to the fibres. Their expressions may be written as follows:

$$E_2^{(2)} = E_2^0 - F_1 D \tag{4.20}$$

and

$$G_{12}^{(2)} = G_{12}^0 - F_2 D \tag{4.21}$$

where F_1 and F_2 are functions of the current values of the elastic moduli and D is a crack density parameter.

Comparing Equation (4.20) with (4.16) and Equation (4.21) with Equation (4.19), we find,

$$F_1 = 2a_{11}\nu_{21}^0 - 2a_5(\nu_{21}^0)^2 - b_4, \tag{4.22}$$

$$F_2 = -\ 1/2\, f_3 \tag{4.23}$$

and

$$D = D_{II}^2 \tag{4.24}$$

Note that the scalar field value D and the vector field value D_{II}^2 are related in accordance with Equation (2.6).

Gottesman et al. have taken $E_1^{(2)} = E_1^0$ and $\nu_{12}^{(2)} = \nu_{12}^0$. Our expressions (4.15) and (4.17) show, however, changes in these moduli with damage. It is likely that these changes may be second-order effects. However, a careful experimental verification of this is required.

Discussion

The vector field characterization of damage presented here appears to have certain inherent properties that would not be expected in a scalar field characterization. For instance, the description of the changes in symmetry properties caused by an array of parallel cracks in a composite laminate, which we have illustrated above, is a consequence of the vector field characterization. A further consequence is the description of the changes in all the elastic moduli caused by various damage modes.

A limitation of the vector field characterization is the fact that crack shape effects cannot be accounted for. A proper account of these effects would be taken by a tensor field characterization that incorporates the crack surface motions. However, in such a case, a larger number of material constants would have to be determined than is the case for the present approach. The usefulness of a given approach would, of course, depend on the feasibility of determining the material constants involved. For the present approach, for instance, the number of unknown constants is four in the case of a single damage field in a two-dimensional orthotropic composite. The experimental determination of these constants is under investigation and will be reported in future.

We should recall here that we have, in the present work, treated the problem of describing the elastic response of a composite following attainment of a

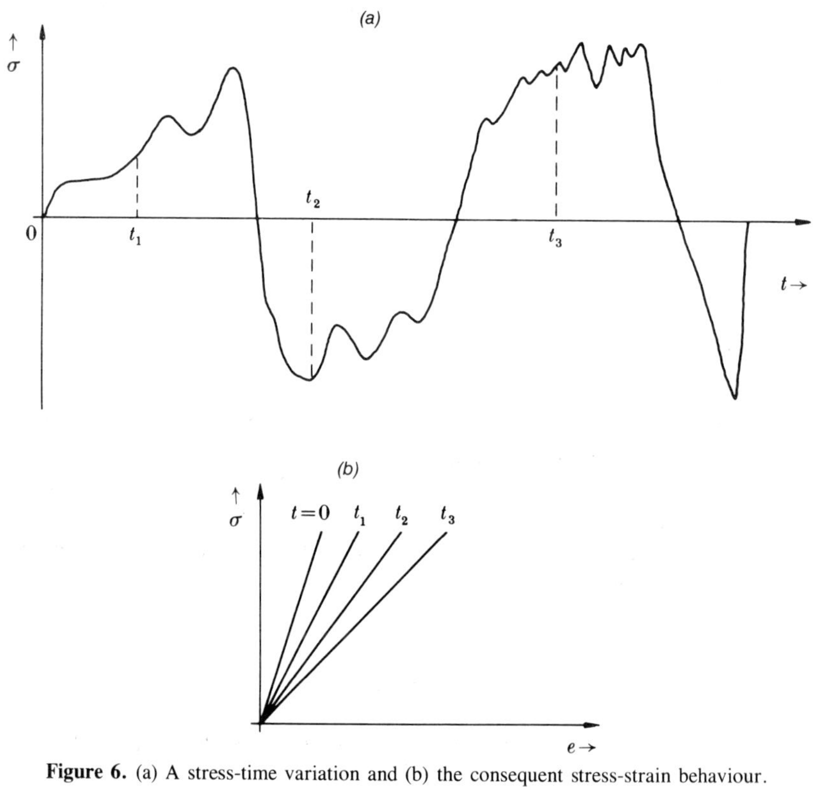

Figure 6. (a) A stress-time variation and (b) the consequent stress-strain behaviour.

damage state. The development of damage under a specified loading history has not been treated. To illustrate this let us suppose that a composite is subjected to a stress-time variation shown in Figure 6a. Figure 6b illustrates the initial stress strain response and the stress-strain response that would be obtained if it were measured following loading to various instants shown in Figure 6a. The present approach would enable us to predict the linear elastic response at these instants if we know the state of damage (i.e. the value of the damage vector field) at these instants. On the other hand, if the elastic moduli are measured at these instants, then, at least for a single orientation of the damage field, we can infer the development of damage from one instant to another. Various investigators (e.g. Highsmith and Reifsnider 1982) have reported elastic moduli of composite laminates that were measured continuously during monotonically increasing and fatigue loadings. Such measurements can be used, in conjunction with equations such as Equations (4.9–4.13), to describe the development of damage.

It is of course desirable to have equations that describe the general nature of the kinetics of damage development in composites. However, the physical mechanisms governing the kinetics must be understood first. The present understanding of damage in composites is not sufficiently advanced for a proper mechanics treatment of the kinetics in general. For instance, little is understood of the mechanisms of initiation and growth of interlaminar cracks away from the free edges in laminates. Until a mechanics treatment of damage kinetics can be made, the equations derived here for residual properties of damaged composites provide useful means for characterizing the development of damage.

Acknowledgement

The present paper was written while the author was visiting the Engineering Science and Mechanics Department, Virginia Polytechnic Institute and State University, Blacksburg, Virginia. The cordial hospitality of the department is greatly appreciated. The author also wishes to thank Professor K. L. Reifsnider for inspiring discussions.

References

1. Adkins, J. E., *Arch. Rat. Mech Anal.*, 4, p. 193 (1960).
2. Aveston, J. and Kelly, A., *J. Mater. Sc.*, 8, pp. 352–362 (1973).
3. Aveston, J. and Kelly, A., *Phil. Trans. R. Soc. Lond.*, A 294, pp. 519–534 (1980).
4. Coleman, B. D. and Gurtin, M. E., *J. Chem. Phys.*, 47, p. 597 (1967).
5. Coleman, B. D. and Noll, W., *Arch. Rat. Mech. Anal.*, 13, p. 167 (1963).

6. Davison, L. and Stevens, A. L., *J. App. Phys.*, 44, p. 667 (1973).
7. Gottesman, T., Hashin, Z. and Brull, M. A., *Advances in Composite Materials*, Proc. ICCM3, (Ed. by A. R. Bunsell et. al.), pp. 749–758, Pergamon, New York (1980).
8. Hahn, H. T., *Composite Materials: Testing and Design* (Fifth Conference) ASTM STP 674, pp. 383–417, American Society for Testing and Materials, Philadelphia (1979).
9. Harlow, D. G. and Phoenix, S. L., *J. Comp. Mats.*, 12, pp. 196 and 314 (1978).
10. Henneke, E. G. and Reifsnider, K. L., *Comp. Tech. Rev.*, 1, p. 8 (1979).
11. Henneke, E. G., Reifsnider, K. L. and Stinchcomb, W. W., *Proc. ARPA/AFML Review of Progress in Quantitative NDE*, AFML-TR-78-205, Rockwell International, Thousand Oaks, California (1979).
12. Highsmith, A., Reifsnider, K. L., *Damage in Composite Materials*, ASTM STP 775 (Ed. by K. L. Reifsnider), American Society for Testing and Materials, Philadelphia, pp. 103–117 (1982).
13. Jamison, R. D., Ph.D. Thesis, Eng. Sci. & Mech. Dept., Virginia Polytechnic Inst. & State University, Blacksburg, Virginia (1982).
14. Kachanov, L. M., *Izv. AN SSR, Otd. Tekhn. Nauk*, 8, p. 26 (1958).
15. Krajcinovic, D. and Fosenka, G. U., *Trans. ASME, J. App. Mech.*, 48, p. 809 (1981).
16. Kratochvil, J. and Dillon, O. W., Jr., *Trans. ASME, J. App. Mech.*, 40, p. 3207 (1969).
17. Leckie, F. A., *Phil. Trans. Roy. Soc.*, London, A288, p. 27 (1978).
18. O'Brien, T. K., Ph.D. Thesis, Eng. Sci. & Mech. Dept., Virginia Polytechnic Inst. & State University, Blacksburg, Virginia (1978).
19. O'Brien, T. K. and Teifsnider, K. L., *J. Test. Eval.*, 5, p. 384 (1977); *J. Comp. Mats.*, 15, p. 55 (1981).
20. Pipkin, A. C. and Rivlin, R. S., *Arch. Rat. Mech. Anal.*, 4, p. 129 (1959).
21. Reifsnider, K. L., *Int. J. Frac.*, 16, p. 563 (1979).
22. Reifsnider, K. L. and Highsmith, A., *Materials, Experimentation and Design in Fatigue*, Westbury House, Surrey, pp. 246–260 (1981).
23. Reifsnider, K. L. and Talug, A., *Proc. Workshop on Composite Materials Research*, Duke University (1978).
24. Rivlin, R. S., *J. Rat. Mech. Anal.*, 4, p. 681 (1955).
25. Rivlin, R. S. and Eriksen, J. L., *J. Rat. Mech. Anal.*, 4, p. 323 (1955).
26. Sih, G. C., Paris, P. C. and Irwin, G. R., *Int. J. Frac. Mech.*, 1, p. 189 (1965).
27. Smith, G. F., *Quart. App. Math.*, __, p. 509 (1982).
28. Smith, G. F. and Rivlin, R. S., *Arch. Rat. Mech. Anal.*, 15, p. 169 (1964).
29. Smith, G. F., Smith, M. M. and Rivlin, R. S., *Arch. Rat. Mech. Anal.*, 12, p. 93 (1963).
30. Stinchcomb, W. W. and Reifsnider, K. L., *Fatigue Mechanisms, Proc. ASTM-NBS-NSF Symp.*, ASTM STP 679 (Ed. by J. J. Fong), pp. 762–787, American Society for Testing and Materials, Philadelphia (1979).
31. Talreja, R., *Advances in Composite Materials*, Proc. ICCM3, A. R. Bunsell et. al., Pergamon, New York (1980); *Proc. Roy, Soc.* London A378, p. 461, Chapter 2 in the present volume.
32. Talug, A. and Reifsnider, K. L., *J. Fibre Sic. Tech.*, 12, p. 201 (1979).
33. Truesdell, C. and Toupin, R. A., *Encyclopedia of Physics*, ed. by S. Flugge, Vol. 3/1, Springer-Verlag, Berlin (1960).
34. Weyl, H., The Classical Groups, Princeton (1946).
35. Wineman, A. S. and Pipkin, A. C., *Arch. Rat. Mech. Anal.*, 17, p. 184 (1964).

Appendix A. Transversely Isotropic Materials

Let us for simplicity consider only one vector field, **V**. The invariants are now given by

$$
\begin{aligned}
I_1 &= e_{11} \\
I_2 &= e_{22} + e_{33} \\
I_3 &= e_{22}^2 + 2e_{23}^2 + e_{33}^2 \\
I_4 &= e_{12}^2 + e_{13}^2 \\
I_5 &= e_{22}e_{12}^2 + 2e_{12}e_{13} + e_{13}^2e_{33} \\
I_6 &= V_1 \\
V_7 &= V_2e_{12} + V_3e_{13} \\
I_8 &= V_2e_{22}e_{12} + V_2e_{23}e_{13} + V_3e_{23}e_{12} + V_3e_{33}e_{13} \\
I_9 &= V_2^2 + V_3^2 \\
I_{10} &= V_2^2e_{22} + 2V_2V_3e_{23} + V_3^2e_{33}
\end{aligned}
\tag{A.1}
$$

The Helmholz free energy for a transversely isotropic material subjected to isothermal, small strains can now be written as follows

$$\varrho\psi = f(I_1,I_2 \ldots,I_{10}) \tag{A.2}$$

where f is a polynomial function.

For small strains and small magnitude of damage vector, which has been normalized, we retain up to second-order terms in e_{ij} and V_k and write the following polynomial expression for f.

$$
\begin{aligned}
\varrho\psi = {} & A_1I_1^2 + A_2I_1I_2 + A_3I_1^2(a_1I_6 + I_6^2) + A_4I_1I_7 \\
& + A_5I_1^2I_9 + A_6I_1I_{10} + A_7I_1I_2(a_2I_6 + I_6^2) \\
& + A_8I_1I_2I_9 + B_1I_2^2 + B_2I_2^2(bI_6 + I_6^2) + B_3I_2I_7 \\
& + B_4I_2^2I_9 + B_5I_2I_{10} + C_1I_3 + C_2I_3(cI_6 + I_6^2) \\
& + C_3I_3I_9 + D_1I_4 + D_2I_4(dI_6 + I_6^2) + D_3I_4I_9 \\
& + E_1I_6I_8 + F_1I_7^2 + G_1I_8 + P_0 + P_1(e_{ij},V_k) + P_2(V_k).
\end{aligned}
\tag{A.3}
$$

where A_i, B_i, C_i, D_i, E_1, F_1, G_1, P_0, a, b, c and d are material constants, P_1 contains only linear terms in e_{ij} and P_2 is quadratic in V_k.

Setting the Helmholz free energy to zero for the unstrained and undamaged material, we get

$$P_0 = 0 \tag{A.4}$$

Applying the condition that an unstrained material of any damage state must be stress free, we have

$$P_1 = 0 \tag{A.5}$$

when Equation (3.9) is used.

A little reflection shows that two damage vector fields of equal magnitude and opposite directions represent equal damage states, i.e. we must have,

$$\varrho\psi(\mathbf{e},\mathbf{V}) = \varrho\psi(\mathbf{e},-\mathbf{V}) \tag{A.6}$$

Using Equation (A.6) and Equation (A.3), we find that the coefficients of all linear terms in V_k must vanish, i.e. we have,

$$a_1 = a_2 = b = c = d = A_4 = B_3 = G_1 = 0 \tag{A.7}$$

Substituting relations (A.1) in Equation (A.3) and using Equation (A.7) we obtain,

$$\begin{aligned}
\varrho\psi = {} & A_1e_{11}^2 + A_2e_{11}(e_{22} + e_{33}) + (B_1 + C_1)e_{22}^2 + 2B_1e_{22}e_{33} \\
& + (B_1 + C_1)e_{33}^2 + D_1e_{13}^2 + 2C_1e_{23}^2 + D_1e_{12}^2 \\
& + e_{11}^2(A_3V_1^2 + A_5V_2^2 + A_5V_3^2) + e_{22}^2(B_2V_1^2 + B_4V_2^2 \\
& + B_4V_3^2 + B_5V_2^2 + C_2V_1^2 + C_3V_2^2 + C_3V_3^2) + e_{33}^2(B_2V_1^2 \\
& + B_4V_2^2 + B_4V_3^2 + B_5V_3^2 + C_2V_1^2 + C_3V_2^2 + C_3V_3^2) \\
& + e_{11}e_{22}(A_6V_2^2 + A_7V_1^2 + A_8V_2^2 + A_8V_3^2) \\
& + e_{11}e_{33}(A_6V_3^2 + A_7V_1^2 + A_8V_2^2 + A_8V_3^2) \\
& + e_{22}e_{33}(2B_2V_1^2 + 2B_4V_2^2 + 2B_4V_3^2 + B_5V_3^2 + B_5V_2^2) \\
& + 2A_6e_{11}e_{23}V_2V_3 + E_1e_{22}e_{12}V_1V_2 + 2B_5e_{22}e_{23}V_2V_3 \\
& + 2B_5e_{33}e_{23}V_2V_3 + E_1e_{33}e_{13}V_1V_3 \\
& + e_{12}^2(D_2V_1^2 + D_3V_2^2 + D_3V_3^2 + F_1V_2^2) \\
& + e_{13}^2(D_2V_1^2 + D_3V_2^2 + D_3V_3^2 + F_1V_3^2) \\
& + 2C_2e_{23}^2V_1^2 + 2C_3e_{23}^2(V_2^2 + V_3^2) + 2F_1e_{12}e_{13}V_2V_3 \\
& + E_1e_{12}e_{23}V_1V_3 + E_1e_{13}e_{23}V_2V_1 + P_2
\end{aligned} \tag{A.8}$$

In Equation (A.8) the first eight terms on the right hand side are independent of the damage vector components. These terms represent the Helmholz free energy of an undamaged material. We may therefore write,

$$\varrho\psi = \varrho\psi^0 + \varrho\psi^1 \tag{A.9}$$

where

$$\begin{aligned}
\varrho\psi^0 = {} & A_1e_{11}^2 + A_2e_{11}(e_{22} + e_{33}) + (B_1 + C_1)e_{22}^2 \\
& + 2B_1e_{22}e_{33} + (B_1 + C_1)e_{33}^2 + D_1e_{13}^2 + 2C_1e_{23}^2 + D_1e_{12}^2
\end{aligned} \tag{A.10}$$

Appendix B. Orthotropic Materials

The most general polynomial expression for the Helmholz free energy, $\varrho\psi$, having up to second order terms in the arguments e_{ij} and V_k is, on using the invariants shown in Equations (3.27), given by,

$$\begin{aligned}\varrho\psi = {} & a_1 I_1^2 + a_2 I_1 I_2 + a_3 I_1 I_3 + a_4 I_1^2 I_8 \\ & + a_5 I_1^2 I_9 + a_6 I_1^2 I_{10} + a_7 I_1 I_{11} + a_8 I_1 I_{12} + a_9 I_1 I_{13} \\ & + a_{10} I_1 I_2 I_8 + a_{11} I_1 I_2 I_9 + a_{12} I_1 I_2 I_{10} \\ & + a_{13} I_1 I_3 I_8 + a_{14} I_1 I_3 I_9 + a_{15} I_1 I_3 I_{10} + b_1 I_2^2 \\ & + b_2 I_2 I_3 + b_3 I_2^2 I_8 + b_4 I_2^2 I_9 + b_5 I_2^2 I_{10} + b_6 I_2 I_{11} \\ & + b_7 I_2 I_{12} + b_8 I_2 I_{13} + b_9 I_2 I_3 I_8 + b_{10} I_2 I_3 I_9 \\ & + b_{11} I_2 I_3 I_{10} + c_1 I_3^2 + c_2 I_3^2 I_8 + c_3 I_3^2 I_9 + c_4 I_3^2 I_{10} \\ & + c_5 I_3 I_{11} + c_6 I_3 I_{12} + c_7 I_3 I_{13} + d_1 I_4 + d_2 I_4 I_8 \\ & + d_3 I_4 I_9 + d_4 I_4 I_{10} + e_1 I_5 + e_2 I_5 I_8 + e_3 I_5 I_9 \\ & + e_4 I_5 I_{10} + f_1 I_6 + f_2 I_6 I_8 + f_3 I_6 I_9 + f_4 I_6 I_{10} \\ & + g_1 I_{14} + g_2 I_{15} + g_3 I_{16} + P_0 + P_1(e_{ij}, V_k) + P_2(V_i) \qquad \text{(B.1)}\end{aligned}$$

Imposing the conditions that the Helmholz free energy is zero for unstrained and undamaged material, and that an unstrained material is stress free, we get

$$P_0 = P_1 = 0 \qquad \text{(B.2)}$$

Substituting Equation (3.27) in Equation (B.1) and writing $\varrho\psi = \varrho\psi^0 + \varrho\psi^1$, we have

$$\begin{aligned}\varrho\psi^0 = {} & a_1 e_{11}^2 + a_2 e_{11} e_{22} + a_3 e_{11} e_{33} + b_1 e_{22}^2 + b_2 e_{22} e_{33} \\ & + c_1 e_{33}^2 + d_1 e_{23}^2 + e_1 e_{13}^2 + f_1 e_{12}^2 \qquad \text{(B.3)}\end{aligned}$$

and

$$\begin{aligned}\varrho\psi^1 = {} & a_4 e_{11}^2 V_1^2 + a_5 e_{11}^2 V_2^2 + a_6 e_{11}^2 V_3^2 + a_7 e_{11} e_{23} V_2 V_3 \\ & + a_8 e_{11} e_{13} V_1 V_3 + a_9 e_{11} e_{12} V_1 V_2 + a_{10} e_{11} e_{22} V_1^2 \\ & + a_{11} e_{11} e_{22} V_2^2 + a_{12} e_{11} e_{22} V_3^2 + a_{13} e_{11} e_{33} V_1^2 \\ & + a_{14} e_{11} e_{33} V_2^2 + a_{15} e_{11} e_{33} V_3^2 + b_3 e_{22}^2 V_1^2 + b_4 e_{22}^2 V_2^2 \\ & + b_5 e_{22}^2 V_3^2 + b_6 e_{22} e_{23} V_2 V_3 + b_7 e_{22} e_{13} V_1 V_3 \\ & + b_8 e_{22} e_{12} V_1 V_2 + b_9 e_{22} e_{23} V_1^2 + b_{10} e_{22} e_{33} V_2^2 \\ & + b_{11} e_{22} e_{33} V_3^2 + c_2 e_{33}^2 V_1^2 + c_3 e_{33}^2 V_2^2 + c_4 e_{33}^2 V_3^2 \\ & + c_5 e_{33} e_{23} V_2 V_3 + c_6 e_{33} e_{13} V_1 V_3 + c_7 e_{33} e_{12} V_1 V_2 \\ & + d_2 e_{23}^2 V_1^2 + d_3 e_{23}^2 V_2^2 + d_4 e_{23}^2 V_3^2 + e_2 e_{13}^2 V_1^2 \\ & + e_3 e_{13}^2 V_2^2 + e_4 e_{13}^2 V_3^2 + f_2 e_{12}^2 V_1^2 + f_3 e_{12}^2 V_2^2 \\ & + f_4 e_{12}^2 V_3^2 + g_1 e_{13} e_{12} V_2 V_3 + g_2 e_{12} e_{23} V_1 V_3 \\ & + g_3 e_{23} e_{13} V_1 V_2 + P_2 \qquad \text{(B.4)}\end{aligned}$$

CHAPTER 7

Residual Stiffness Properties of Cracked Composite Laminates

Abstract: Cracks in composite laminates are characterized by a set of vectors, each representing an individual cracking mode. The vector components are taken as internal variables in the elastic strain energy function and the elastic constitutive equations are derived for the inplane loading condition. For low concentration of cracks in laminates, the residual stiffness properties are related to the initial elastic constants and the magnitude of the damage vectors. These equations are then used to predict stiffness reductions of composite laminates from the observed crack densities. Very good agreement with the experimental values is found.

Keywords: Composite materials, stiffness changes, matrix cracking, crack density, damage mechanics.

DAMAGE IN COMPOSITE materials is generally of a highly complex nature. However, extensive work done in the last decade on selected configurations of composite laminates using various nondestructive techniques has helped advance the understanding of damage (Hahn, 1979; Reifsnider, 1979; Stinchcomb and Reifsnider, 1979). Perhaps the best understood damage mode is the transverse cracking of off-axis plies in a laminate. Various investigations, e.g. Masters and Reifsnider (1982), have shown that, both under static loads and during initial stages of cyclic loading, multiple cracking of off-axis plies occurs in the form of matrix cracks growing parallel to fibres. The cracks increase in number with increasing load, or with increasing number of cycles of a cyclic load, and a state of saturation eventually ensues. The saturation state has some characteristic features, and has thus been called a characteristic damage state, CDS (Reifsnider and Talug, 1978). For instance, the saturation spacing of cracks depends on the laminate configuration (i.e. fiber orientation and lamina thickness and stacking sequence) but is independent of the loading history. The saturation crack spac-

Reprinted by permission of Pergamon Press, from *Advances in Fracture Research*, *4*, S. R. Valluri, et al., eds. (1984).

ing has been predicted for various laminates using a shear-lag model (Reifsnider, Henneke and Stinchcomb, 1979).

The reduction in stiffness of laminates undergoing transverse cracking in off-axis plies has been studied both experimentally and analytically (Highsmith and Reifsnider, 1982). The prediction of stiffness has been based on the one-dimensional shear-lag model (Reifsnider, Henneke and Stinchcomb, 1979), which determines the normal stress in the given direction in the cracked plies approximately. For a given crack spacing in a ply, then, the elastic modulus of the ply normal to the cracks is reduced by an amount corresponding to the reduction in the contribution of the cracked ply to the overall load carried by the laminate. The laminate analysis is then used to calculate the overall stiffness constants of the laminate using the degraded modulus of the cracked ply. The predictions agree generally with the experimental values, although occasional discrepancies are found. However, it is not clear to what extent the simplifying assumptions in the analysis affect predictions.

We shall here present a procedure for predicting stiffness reduction due to cracking using an entirely different approach. Rather than considering cracks explicitly in terms of their effects on stress redistributions, we shall look at cracks as microstructural entities such as dislocations and voids in metals. The presence of cracks will then be accounted for by introducing a set of internal state variables in the mechanical response functions. The approach along these lines has been developed for a general case of damage in composites in Chapter 6; we shall here specialize it for transverse cracking of off-axis plies in laminates. Specific equations relating the intensity of cracking and stiffness reductions will then be given for selected laminate configurations and predictions using these equations will be compared with experimentally observed values.

Elastic Response of Cracked Laminates

Consider a solid containing m sets of parallel planar cracks. Let us assign a vector $\mathbf{V}^{(\alpha)}$ to each of the set of cracks, $\alpha = 1,2, \ldots m$. Let the orientation of a vector $\mathbf{V}^{(\alpha)}$ be normal to the planes of the αth set of cracks and let its magnitude be given by

$$|\mathbf{V}^{(\alpha)}|^2 = \eta_c \bullet \bar{\ell}_c \bullet \bar{w}_c \bullet f_c, \tag{1}$$

where η_c is the crack number density, i.e. the number of cracks per unit surface area, $\bar{\ell}_c$ and $\bar{w}_c$ are the average length and the average width of the cracks, respectively, and f_c is a factor that depends on the crack shape and on the constraint to crack opening provided by the surrounding material.

Figure 1 illustrates the case of two sets of parallel cracks in a solid and shows the two vectors $\mathbf{V}^{(1)}$ and $\mathbf{V}^{(2)}$ that represent the two cracking modes.

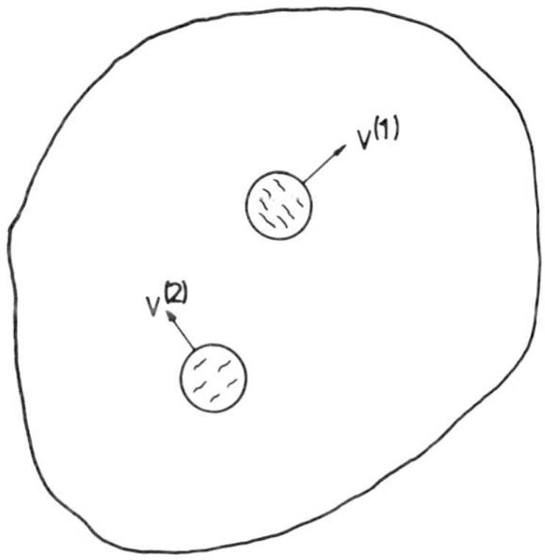

Figure 1. A solid containing two sets of parallel cracks.

The magnitudes of vectors can vary over the solid to account for inhomogeneous cracking modes.

Consider now cracked composite laminates with following restrictions.

1. Thickness is small,
2. loading is in the plane of laminates, and
3. each cracking mode consists of transverse cracking in the plies only.

We thus do not consider other cracking modes such as delamination, fiber breakage and interfacial debonding.

The transverse cracking will now be represented by vectors $\mathbf{V}^{(\alpha)}$, each vector oriented normal to fibres in a cracked ply. Let $V_p^{(\alpha)}$ be the vector components in a Cartesian coordinate system placed in the midplane of the laminate. Considering the vector components as internal variables representing the micro(crack)structure, we write the elastic strain energy function for the cracked laminate as

$$W = W(e_p, V_q^{(\alpha)}) \tag{2}$$

where e_p are the strain tensor components in the Voigt notation with $p = 1,2$ and 6.

For simplicity, let us consider one vector $\mathbf{V}$ representing a single cracking mode in a laminate. Expressing the strain energy function as a polynomial P, we have

$$W = P(e_1, e_2, e_6, V_1, V_2) \tag{3}$$

The form of the polynomial P will be restricted by the symmetry properties of a laminate. The restrictions are accounted for by taking the polynomial P as a function of the integrity bases (invariants) associated with the given symmetry. For a polynomial, which is a function of a symmetric second-order tensor and a vector, the integrity bases associated with orthotropic symmetry have been given by Smith, Smith and Rivlin (1963). Using these, we have the following invariants for our case.

$$I_1 = e_1,\ I_2 = e_2,\ I_3 = e_6^2$$
$$I_4 = V_1^2,\ I_5 = V_2^2,\ I_6 = V_1 V_2 e_6 \tag{4}$$

The most general polynomial P restricted to quadratic terms in strain components and quadratic terms in vector components is now given by

$$\begin{aligned} W = {} & k_1 I_1^2 + k_2 I_1 I_2 + k_3 I_1^2 I_4 + k_4 I_1^2 I_5 + k_5 I_1 I_6 + k_6 I_2^2 + k_7 I_2^2 I_4 \\ & + k_8 I_2^2 I_5 + k_9 I_2 I_6 + k_{10} I_3 + k_{11} I_3 I_4 + k_{12} I_3 I_5 + k_{13} I_1 I_2 I_4 \\ & + k_{14} I_1 I_2 I_5 \\ & + P_0 + P_1(e_p, V_q) + P_2(V_p), \end{aligned} \tag{5}$$

where P_0 is a constant and P_1 and P_2 are linear and quadratic functions of their variables, respectively.

Using (4) in (5) we get,

$$\begin{aligned} W = {} & k_1 e_1^2 + k_2 e_1 e_2 + k_3 e_1^2 V_1^2 + k_4 e_1^2 V_2^2 + k_5 e_1 e_6 V_1 V_2 + k_6 e_2^2 \\ & + k_7 e_2^2 V_1^2 + k_8 e_2^2 V_2^2 + k_9 e_2 e_6 V_1 V_2 + k_{10} e_6^2 + k_{11} e_6^2 V_1^2 + k_{12} e_6^2 V_2^2 \\ & + k_{13} e_1 e_2 V_1^2 + k_{14} e_1 e_2 V_2^2 + P_0 + P_1 + P_2 \end{aligned} \tag{6}$$

Imposing the condition,

$$W = 0 \text{ for } e_p = 0 \text{ and } V_q = 0, \text{ we get,}$$
$$P_0 = 0 \tag{7}$$

and further,

$$\sigma_p = \frac{\partial W}{\partial e_p} = 0 \text{ for } e_p = 0 \text{ gives}$$
$$P_1 = 0 \tag{8}$$

Applying the stress formula we now obtain,

$$\sigma_p = C_{pq} e_q = (C_{pq}^0 + C^1{}_{pq}) e_q \tag{9}$$

where

$$C^0_{pq} = \begin{bmatrix} 2k_1 & k_2 & 0 \\ k_2 & 2k_6 & 0 \\ 0 & 0 & 2k_{10} \end{bmatrix} \tag{10}$$

and

$$C^1_{pq} = \begin{bmatrix} (2k_3V_1^2 + 2k_4V_2^2) & (k_{13}V_1^2 + k_{14}V_2^2) & k_5V_1V_2 \\ (k_{13}V_1^2 + k_{14}V_2^2) & (2k_7V_1^2 + 2k_8V_2^2) & k_9V_1V_2 \\ k_5V_1V_2 & k_9V_1V_2 & (2k_{11}V_1^2 + 2k_{12}V_2^2) \end{bmatrix} \tag{11}$$

C^0_{pq} is the orthotropic stiffness matrix of the uncracked laminate and C^1_{pq} corresponds to the contribution due to cracks. Equation (11) shows that cracking will, in general, remove the initial orthotropic symmetry in a laminate.

Residual Stiffness Properties

The stiffness components of a cracked laminate with transverse cracking in one ply are given by

$$C_{pq} = C^0_{pq} + C^1_{pq} \tag{12}$$

It can be shown that for cracking in m plies the stiffness components are given by (Talreja, 1983)

$$C_{pq} = C^0_{pq} + C^\alpha_{pq} \tag{13}$$

where $\alpha = 1,2,3, \ldots m$.

Let us first consider one cracking mode. Figure 2 shows a cross-ply laminate which has cracks in the 90° plies. The damage vector for this case is given by

$$\mathbf{V} = D(1,0) \tag{14}$$

Substituting this in Equation (11) and using Equation (12), we obtain

$$C_{pq} = \begin{bmatrix} (2k_1 + 2k_3D^2) & (k_2 + k_{13}D^2) & 0 \\ (k_2 + k_{13}D^2) & (2k_6 + 2k_7D^2) & 0 \\ 0 & 0 & (2k_{10} + 2k_{11}D^2) \end{bmatrix} \tag{15}$$

The orthotropic symmetry in the stiffness coefficients is thus retained for this cracking mode.

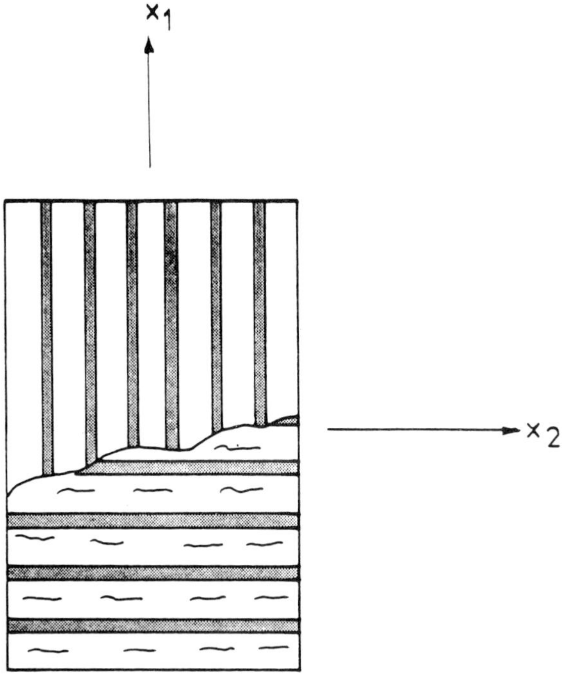

Figure 2. A cross-ply laminate with cracks in the 90° ply.

The residual elastic moduli can now be calculated by using the relations:

$$E_1 = \frac{C_{11}C_{22} - C_{12}^2}{C_{22}}, \; E_2 = \frac{C_{11}C_{22} - C_{12}^2}{C_{11}}$$
$$\nu_{12} = \frac{C_{12}}{C_{22}} \text{ and } C_{12} = C_{66} \tag{16}$$

From Equations (15) and (16) we obtain,

$$E_1 = E_1^0 + 2D^2[k_3 + k_7(\nu_{12}^0)^2 - k_{13}\nu_{12}^0]$$
$$E_2 = E_2^0 + 2D^2[k_7 + k_3(\nu_{21}^0)^2 - k_{13}\nu_{21}^0] \tag{17}$$
$$\nu_{12} = \nu_{12}^0 + D^2\left[\frac{1 - \nu_{12}^0\nu_{21}^0}{E_2^0}\right](k_{13} - 2k_7\nu_{12}^0) \text{ and } G_{12} = G_{12}^0 + 2D^2K_{11}$$

In Equations (17) the moduli with superscript 0 refer to the uncracked laminate.

Next, let us consider two cracking modes. Figure 3 shows a laminate with cracks in two off-axis plies placed symmetrically about the axis x_1. The damage vectors in this case are

$$\mathbf{V}^{(1)} = D_1(\sin\theta, \; -\cos\theta) \tag{18}$$

and

$$\mathbf{V}^{(2)} = D_2(\sin\theta, \cos\theta) \tag{19}$$

where θ is the off-axis angle.

For $\theta = 45°$ the stiffness matrix is given by

$$C_{pq} = \begin{bmatrix} [2k_1 + 2k_3(D_1^2 + D_2^2)] & [k_2 + k_{13}(D_1^2 + D_2^2)] & [1/4 \; k_5(D_2^2 - D_1^2)] \\ & [2k_6 + 2k_7(D_1^2 + D_2^2)] & [1/4 \; k_9(D_2^2 - D_1^2)] \\ & & [2k_{10} + 2k_{11}(D_1^2 + D_2^2)] \end{bmatrix} \tag{20}$$

where only half of the symmetric matrix is shown.

It is seen that, for $D_1 = D_2$, the orthotropic symmetry is retained. However, we shall assume that, even for $D_1 \neq D_2$, the orthotropic symmetry

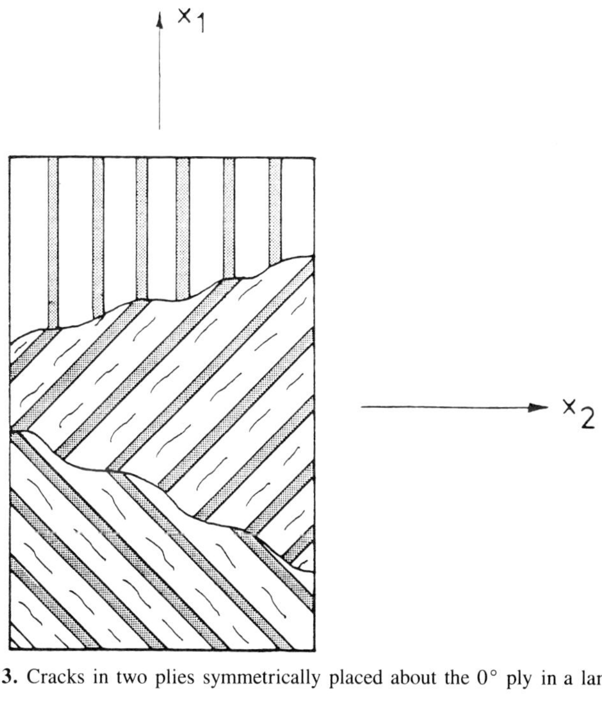

Figure 3. Cracks in two plies symmetrically placed about the 0° ply in a laminate.

will be retained approximately. Equivalently, we shall assume that the constants k_5 and k_9, which represent the interactions between the normal and the shear strains in a cracked laminate, are small.

The residual elastic moduli for this case are given by

$$E_1 = E_1^0 + 2(D_1^2 + D_2^2)[k_3 + k_7(\nu_{12}^0)^2 - k_{13}\nu_{12}^0]$$

$$E_2 = E_2^0 + 2(D_1^2 + D_2^2)[k_7 + k_3(\nu_{21}^0)^2 - k_{13}\nu_{21}^0] \tag{21}$$

$$\nu_{12} = \nu_{12}^0 + (D_1^2 + D_2^2)\left[\frac{1 - \nu_{12}^0\nu_{21}^0}{E_2^0}\right](k_{13} - 2k_7\nu_{12}^0)$$

and

$$G_{12} = G_{12}^0 + 2(D_1^2 + D_2^2)k_{11}$$

Prediction of Stiffness Reductions

Highsmith and Reifsnider (1982) have reported stiffness reductions and crack densities for the following 4 laminates of glass/epoxy.

Laminate 1: $(0,90_3)_s$, Laminate 2: $(90_3,0)_s$

Laminate 3: $(0,90)_s$, Laminate 4: $(0,\pm 45)_s$

We shall use the data for laminate 1 to calculate the unknown constants in the stiffness reduction Equation (17) and use these to predict the stiffness reductions for the remaining laminates. Since the measurement of the shear modulus G_{12} is suspect due to the method used (Highsmith and Reifsnider, 1982), we shall not concern ourselves with predicting this property. Also, since the change in the shear modulus is independent of the changes in the other moduli for the cracking modes in the laminates considered, it should suffice to treat the other moduli for the sake of demonstrating the prediction capability of the method.

The basic data for a single ply and the initial elastic moduli of the laminates, calculated by laminate analysis, are shown in Table 1.

The squares of the magnitudes of damage vectors, D^2, are calculated using the definition given by Equation (1) for the cracking modes in all four laminates, and are given in Table 2. Also shown in this table are the observed values of the number of cracks per unit length of specimen. The following procedure is used in calculating D^2.

The average crack length, $\bar{\ell}_c$ is taken as the width of a specimen W for cracks in 90° plies, since the observation shows that the cracks span the entire dis-

Table 1. Elastic moduli for glass/epoxy lamina and laminates. Single ply thickness = 0.203 mm.

Moduli	E_1, GPa	E_2, GPa	ν_{12}	G_{12}, GPa
Lamina	41.7	13.0	0.300	3.4
Laminate 1	20.3	34.75	0.112	—
Laminate 2	20.3	34.75	0.112	—
Laminate 3	27.57	27.57	0.142	—
Laminate 4	21.79	13.93	0.561	—

tance from edge to edge of a specimen soon after initiation (Highsmith and Reifsnider, 1982). For the 45° plies the crack length is taken as W/cos 45°. The average width of cracks, $\bar{w}_c$ is equated to the thickness of the plies that contain the cracks, e.g. for laminate 1, $\bar{w}_c = 6t_0$, where t_0 is the single ply thickness. The crack number density, η_c is equated to the observed number of cracks per unit length of specimen divided by the width W.

The factor f_c is assumed to be a function of the crack width to the specimen thickness ratio and is taken as

$$f_c = \frac{nt_0}{t}, \tag{22}$$

where n is the number of plies containing a transverse crack and t is the thickness of the specimen (laminate). Due to the lack of information regarding the nature of the constraint on crack opening and sliding, any further refinement of this factor is not possible at present. We have further assumed that for cracks ending in the free surface, e.g. in $(90_3,0)_s$ laminate, the constraint is essentially absent, i.e. $f_c = 1$.

Table 2. Magnitudes of the damage vectors.

	$n_c \cdot w$ mm^{-1}	$\bar{\ell}_c$ mm	$\bar{w}_c$ mm	f_c	D^2
Laminate 1. Cracks in 6 90° plies	0.75	W	1.218	0.75	0.6851
Laminate 2. Cracks in 3 90° plies	0.51	W	0.609	1.0	0.3106
Laminate 3. Cracks in 2 90° plies	1.90	W	0.406	0.50	0.3857
Laminate 4. Cracks in +45° ply	1.27	1.41W	0.203	0.17	0.0609
Cracks in −45° ply	0.85	1.41W	0.203	0.17	0.0406

Table 3. Predicted and observed stiffness reduction of cracked laminates.

	ΔE_1, %		$\Delta \nu_{12}$, %		ΔE_2, %	
Laminate	Observed	Predicted	Observed	Predicted	Observed	Predicted
2	38.8	38.1	71.0	67.4	0.0	0.5
3	17.1	17.2	49.8	41.4	7.3	0.4
4	9.4	8.3	8.1	7.5	5.3	0.1

$k_3 = -6.713$ GPa, $k_7 = -0.762$ GPa and $k_{13} = -4.467$ GPa. Data for Laminate 1: $\Delta E_1 = 42.0\%$, $\Delta \nu_{12} = 74.3\%$ and $\Delta E_2 = 0.6\%$.

Table 3 shows the three constants found by using the data for laminate 1, and the predictions of stiffness reductions using these for the other laminates. The observed values, as reported in Highsmith and Reifsnider (1982), are also shown. The accuracy of the predictions may be characterized as very good.

References

1. Hahn, H. T., "Fatigue Behavior and Life Prediction of Composite Laminates," *Composite Materials: Testing and Design*, (Fifth Conf.), *ASTM STP 674*, pp. 383–417 (1979).
2. Highsmith, A. L. and Reifsnider, K. L., "Stiffness-Reduction Mechanisms in Composite Laminates," *Damage in Composite Materials*, *ASTM STP 775*, pp. 103–117 (1982).
3. Masters, J. E. and Reifsnider, K. L., "An Investigation of Cumulative Damage Development in Quasi-Isotropic Graphite/Epoxy Laminates," *Damage in Composite Materials*, *ASTM STP 775*, pp. 40–62 (1982).
4. Reifsnider, K. L., "Fatigue Behavior of Composite Materials," *Int. J. of Fracture*, *16*, No. 6, pp. 563–583 (1979).
5. Reifsnider, K. L., Henneke, E. G. and Stinchcomb, W. W., "Defect-Property Relationships in Composite Materials," *AFML-TR-76-81*, Part IV, Air Force Materials Laboratory.
6. Reifsnider, K. L. and Talug, A., "Characteristic Damage States in Composite Laminates," *Proc. of Workshop on Composite Materials Research*, Duke University (October 17–18, 1978).
7. Smith, G. F., Smith, M. M. and Rivlin, R. S., "Integrity Bases for a Symmetric Tensor and a Vector—The Crystal Classes," *Arch. Rational Mech. Anal.*, Vol. 12, pp. 93–133 (1963).
8. Stinchcomb, W. W. and Reifsnider, K. L., "Fatigue Damage Mechanisms in Composite Materials: A Review," *Fatigue Mechanisms*, *Proc. of ASTM-NBS-NSF Symp.*, Kansas City, Mo., *ASTM STP*, pp. 762–787 (1979).
9. Talreja, R., "A Continuum Mechanics Characterization of Damage in Composite Materials," *The Danish Center for Applied Mathematics and Mechanics*, Report No. 268, Chapter 6 in the present volume (1983).

CHAPTER 8

Transverse Cracking and Stiffness Reduction in Composite Laminates

Abstract: A systematic classification of the effects of transverse cracking on the stress-strain response of composite laminates is presented. Stiffness reductions resulting from transverse cracking in glass/epoxy and graphite/epoxy laminates from crack initiation to crack saturation are predicted using the stiffness-damage relationships developed by the author in a previous work. Good agreement with the experimental results is found. An assessment of the ply discount method for predicting stiffness reductions at crack saturation is also made.

MATRIX CRACKING IN off-axis plies of laminated fibrous composites referred to as transverse cracking, has been a subject of extensive research for over a decade. Since the early observation of the cracking in cross-ply laminates, e.g. by Broutman and Sahu [1], several studies have investigated the conditions surrounding the formation of the cracks. Most studies have been concerned with determining the threshold stress (or strain) for the onset of cracking and the spacing between the cracks. Garrett and Bailey [2], for instance, reported experimental measurements of the threshold stress and the crack spacings for a cross-ply laminate of glass/polyester. They also modified the multiple fracture theory advanced by Aveston and Kelly [3] for unidirectional composites to predict the crack spacings. Later, Parvizi, Garrett and Bailey [4] studied the constraint to transverse cracking in the inner 90° plies provided by the outer 0° plies in glass/epoxy cross-ply laminates and showed that the threshold strain for transverse cracking depended on the thickness of the plies undergoing cracking. Indeed, the onset of transverse cracking could be suppressed beyond the laminate failure by sufficiently reducing the thickness of the inner plies. Similar results were presented by Bader, et al. [5] for graphite/epoxy cross-ply laminates. The observed constraint effect could be accounted for, at least qualitatively, by the multiple fracture theory of Aveston and Kelly [3].

The Aveston-Kelly theory analyzes the energy balance associated with the crack initiation but does not consider the crack growth process. Wang and

Crossman [6] assumed, however, a matrix microcrack to exist and analyzed the conditions for its growth. They calculated the energy release rate associated with the crack growth using a finite element crack closure technique. Assuming that the crack growth would become unstable when the energy release rate reaches the critical value for the material, they calculated the threshold stress. The distance between two adjacent cracks below which the cracks interact and reduce the energy release rate was taken as the crack saturation spacing. The analysis was tested on transverse cracking in 90° plies constrained by ±25° plies and good agreement was found [7].

Reifsnider, et al. [8] investigated transverse cracking in multidirectional laminates and discovered the existence of a crack pattern in a given laminate which they described as a Characteristic Damage State, CDS. The crack pattern was found to develop at sufficient loading, i.e. at a sufficiently high monotonic load or at a sufficiently large number of cycles of a given maximum load, and remain stable on further loading. The CDS consisted of nearly uniform spacings between cracks in individual off-axis plies. The crack spacings were found to depend on the laminate configuration i.e. ply thicknesses, fiber orientations, stacking sequence, etc. The distinctive crack patterns could be predicted by a shear lag analysis [9].

The initiation of cracks and the multiplication of cracks to saturation (CDS) are the important physical aspects of the transverse cracking process which have been clarified by the studies cited above. However, the consequences of transverse cracking on the materials response, as exemplified by changes in the stiffness, strength and fatigue properties, are of major importance to the design of composite structures. We shall consider the stiffness property as affected by the transverse cracking in this work. The characteristics of the stress-strain response reflecting the development of transverse cracking will be described and classified on the basis of the constraint to transverse cracking. The method for predicting stiffness changes caused by transverse cracking based on a continuum damage characterization developed in Chapters 6 and 7, will then be described. The changes in the longitudinal elastic moduli from crack initiation to crack saturation will be predicted and compared with experimental results for various glass/epoxy and graphite/epoxy laminates. The approximate method in use for predicting stiffness loss at crack saturation, called the ply discount method, will be assessed on the basis of the available data.

Transverse Cracking Effects on Stress-Strain Response

Consider a laminate consisting of plies of various fiber orientations subjected to a tensile stress in the longitudinal direction. Under a stress exceeding the threshold stress, one or more of the off-axis plies may develop cracks along the fibers in the plies. We shall call these plies "cracking plies" and the remain-

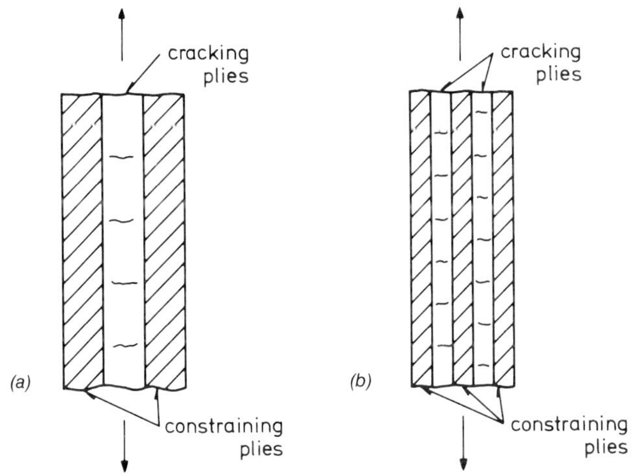

Figure 1. Edge view of laminates undergoing cracking: a) cracking plies are lumped together, b) cracking plies are distributed.

ing plies "constraining plies." The mixing of the two types of plies in a laminate may be in a lumped manner (Figure 1a) or in a distributed manner (Figure 1b).

A remote stress on a cracked laminate would induce crack surface displacements, i.e. opening and sliding, which in turn would introduce non-uniform strains in the constraining plies. The cracking plies would undergo stress relaxation locally in the vicinity of cracks while the constraining plies would carry enhanced stresses in these regions. The local stress redistribution between the plies could in general be of a complex nature, as exemplified by numerical results of Talug [12] for two quasi-isotropic laminates. An example of the strain enhancement due to cracking was obtained by Highsmith [13] who measured surface strains on a $(0,90_3)_s$ laminate by Moiré interferometry. The uniaxial strain variation in a cracked laminate is shown schematically in Figure 2 on the basis of numerical and experimental studies. It is clear that the overall strain averaged over a sufficiently large gauge length would be higher than the uniform strain ϵ_o for the uncracked laminate. The additional average strain for a given remote stress is the underlying cause of the stiffness loss. In a general case, all strain components would be affected leading to changes in all elastic moduli of a laminate.

The initiation and the development of transverse cracking are subject to constraints that come from two sources: one, the thickness of cracking plies which sets to the size of the transverse crack the limit which it can attain before being arrested by the constraining plies, and two, the stiffness of the constraining plies relative to the stiffness of the cracking plies. The constraint effect due to thickness of cracking plies as reflected in the strain to onset of transverse cracking was demonstrated by Baider, et al. [5]. Their data are

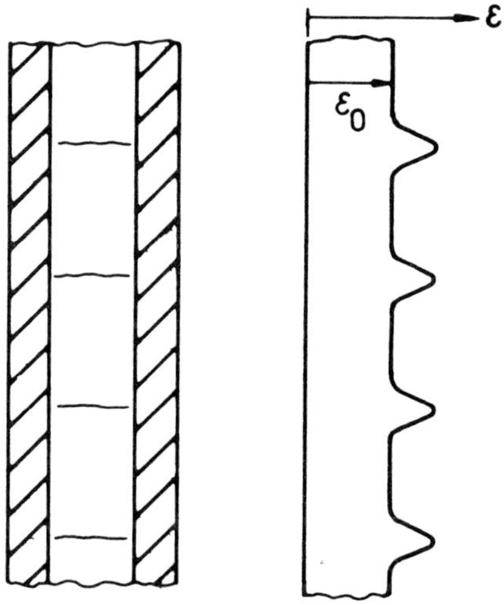

Figure 2. Longitudinal strain variation on the surface of a cracked laminate.

shown in Figure 3 where the strain to onset of transverse cracking denoted by ϵ_{FPF} is plotted against $2n$, the number of transverse plies in a $(0_4, 90_n)_s$ graphite/epoxy laminate. The failure strain for the longitudinal plies is also shown in the figure. The constraint effect related to the stiffness of the constraining plies was studied by Flaggs and Kural [14] for the class of $(\pm\theta,90)_s$ laminates, and it was found that increasing stiffness of the constraining plies in the direction normal to the crack planes led to increased strain to onset of cracking.

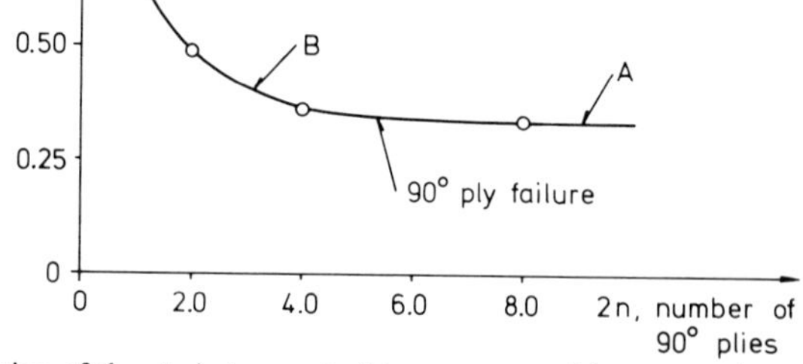

Figure 3. Variation of the strain to onset of transverse cracking, ϵ_{FPF}, along the thickness of the transverse plies.

As a result of the effect of the constraints on the development of transverse cracking, the characteristics of the deformation response would depend on the degree of constraint. For a qualitative characterization of the deformation response, it may be useful to classify the degree of constraint into the following four types.

Type A: no constraint
Type B: low constraint
Type C: high constraint
and Type D: full constraint

The regions of the curve in Figure 3 representing transverse cracking are marked by the appropriate letters in accordance with the proposed classification. Obviously there is no sharp border between the intermediate cases, type B and type C, while the extreme cases, type A and type D represent limiting conditions.

Figure 4 shows the stress-strain behavior corresponding to the limiting case of no constraint, type A. The initial linear elastic response ends when the longitudinal strain reaches the strain to unconstrained failure of the cracking plies. The constraining plies, which provide no constraint in this case, are too weak to carry the additional load shed by the cracking plies and consequently extend instantaneously to failure when the strain reaches their failure strain. The cracking plies would suffer a single fracture, i.e. failure from a single crack, as the (non)constraining plies would be unable to transfer the additional stress back to the cracking plies. The stress-strain behavior displayed here is seemingly that of an elastic-perfect plastic material. However, the unloading

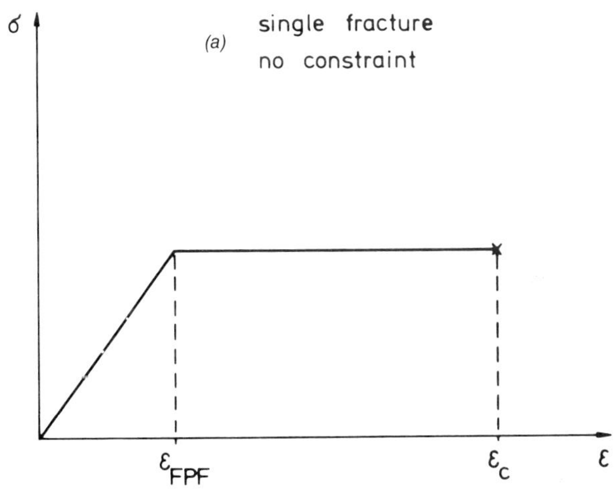

Figure 4. Schematic stress-strain diagram for type A transverse cracking.

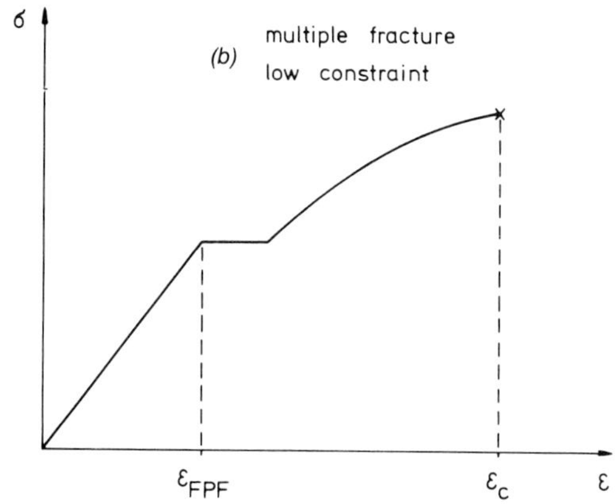

Figure 5. Schematic stress-strain diagram for type B transverse cracking.

path would be along a straight line ending in the origin, as the crack surface displacements are relaxed upon removal of stress.

Figure 5 shows the stress-strain behavior for low constraint, type B. The onset of cracking occurs at a slightly higher strain than in the previous case, at which point multiple parallel cracks appear leading to an abrupt increase in strain. On attainment of the energy balance, the cracking process stops at the load applied and further cracking then requires an increasing load, as discussed by Aveston and Kelly [14]. The stable crack development under increasing load results in the nonlinear stress-strain behavior.

The stress-strain behavior characteristic of the high constraint, type C, is shown in Figure 6. The strain to onset of transverse cracking is enhanced significantly and the additional strain due to cracking is low. The development of cracking requires load increment resulting in a stable crack multiplication. The low strain increase due to cracking gives negligible nonlinearity, and the initiation of cracking is represented by a knee in the stress-strain curve.

Figure 7 shows the limiting case of full constraint, type D, which results in total suppression of transverse cracking. The laminate failure thus results when the failure strain of the constraining plies is reached. The stress-strain behavior remains unaffected.

The uniaxial stress-strain behavior described above for the different cases serves to illustrate the effects of transverse cracking on the deformation response of laminates. The basic characteristics of the constraint effects would, however, apply to general multi-axial stress states.

Stiffness-Transverse Cracking Relationships

The first work to associate stiffness degradation with transverse cracking was by Hahn and Tsai [15], who considered uniaxial behavior of cross-ply

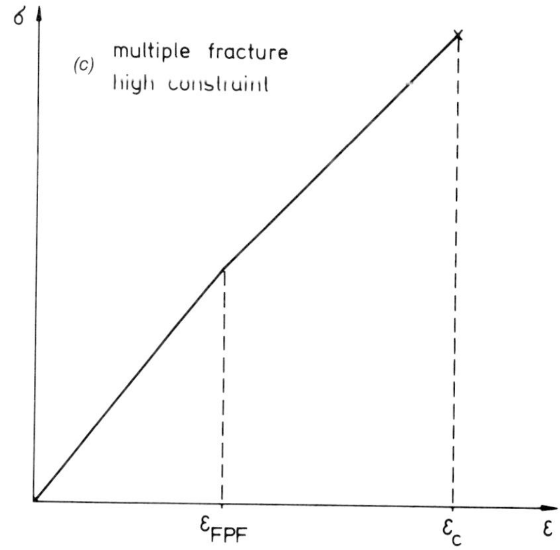

Figure 6. Schematic stress-strain diagram for type C transverse cracking.

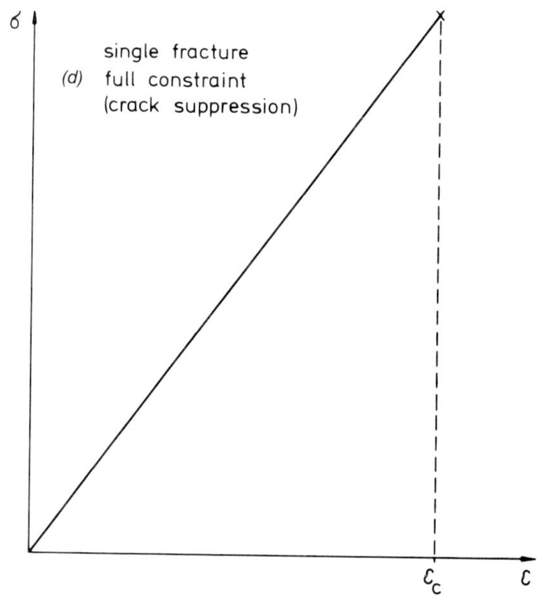

Figure 7. Schematic stress-strain diagram for type D transverse cracking.

laminates. Assuming cracking in the 90° plies to result in reduction of the cross-section locally, they derived expressions for the overall longitudinal modulus that showed bilinear stress-strain behavior. Their model corresponds to the high constraint transverse cracking case (type C, in our classification) and provides, for this case, simple means for assessing the extent of transverse cracking through reduction in the longitudinal modulus.

Gottesman, Hashin and Brull [16] considered a lamina in plane stress condition having cracks parallel to the fibers. Assuming the cracks to be non-interacting and using the expressions for the elastic energy associated with a single crack in an infinite orthotropic sheet, they derived expressions for the overall elastic constants of a cracked lamina. They also gave the estimates for the effective elastic constants by the self-consistent scheme and gave the bounds to these constants by a variational method. A basis for estimating stiffness changes due to transverse cracking for a two-dimensional case was thus established by this work. However, an experimental verification of the accuracy of this method remains to be made.

Recently, Laws, Dvorak and Hejazi [17] presented a different method for estimating stiffness reductions in a cracked lamina. They assumed the elastic moduli of a cracked lamina to be given by a similarly cracked infinite medium with the same fiber volume fraction and the same crack density. The overall elastic moduli of the infinite medium were estimated by a self-consistent method assuming the medium to have three phases: matrix, fibers and cracks that were taken as a limiting case of elliptical voids. This approach is more general than that of Gottesman, Hashin and Brull in the sense that it is three-dimensional. However, the nature of assumptions involved, especially as regards crack geometry, would require a thorough experimental verification.

The present author presented a damage mechanics approach to describing the elastic constitutive behavior of composite materials having a finite number of damage modes, e.g. sets of parallel cracks in Chapter 6. The theory characterized each oriented damage mode by a vector field and derived relationships between the overall elastic moduli and magnitudes of the damage vectors. In addition to providing a basis for calculating stiffness changes from crack densities, the theory showed, for the first time, how symmetry properties of composite materials may be affected by damage modes. The theory was applied to orthotropic laminates in plane stress undergoing transverse cracking in Chapter 7 and a method for predicting elastic moduli changes was presented and applied to the data reported by Highsmith and Reifsnider [18] with satisfactory results.

In the present work, we shall utilize the stiffness-damage relationships developed earlier in these two chapters. For this purpose we shall extract the substance of relevance from these references and summarize it below.

Consider a two-dimensional solid containing m sets of oriented crack arrays. To each parallel crack array, we assign a vector $\mathbf{V}$ oriented normal to the crack planes as illustrated in Figure 8 for two crack arrays. We thus have

$$\mathbf{V}^{(\alpha)} = D^{(\alpha)}\mathbf{n}^{(\alpha)}, \quad \alpha = 1,2, \ldots m \tag{1}$$

where $\mathbf{n}^{(\alpha)}$ is a unit vector oriented normal to the planes of the αth crack array and $D^{(\alpha)}$ is the magnitude of the vector $\mathbf{V}^{(\alpha)}$.

For a given crack array, we take the vector magnitude given by,

$$D^2 = \mathbf{V} \cdot \mathbf{V} = \eta_c \cdot \bar{\ell}_c \cdot \bar{w}_c \cdot f_c \tag{2}$$

where η_c is the number of cracks per unit surface area and $\bar{\ell}_c$ and $\bar{w}_c$ are the average length and width, respectively, of the cracks in the array. The factor f_c is a nondimensional factor taken to account for the constraint to the crack surface displacements. This constraint, as discussed in the previous section, is due to the finite thickness of the cracking plies and due to the stiffness of the constraining plies. We include only the thickness effect in the constraint factor in a rather simplistic way by taking this factor as

$$f_c = \frac{t_c}{t}, \tag{3}$$

where t_c is the thickness of the ply (or adjacent plies) containing the crack array and t is the total thickness of laminate.

Assuming now that the energy function is a function of the strain tensor and the damage vector set $\mathbf{V}^{(\alpha)}$, and assuming further that the strain components and the damage vector magnitudes are small, we can derive the elastic constitutive equations for a composite laminate expressed as

$$\sigma_p = C_{pq}\epsilon_q = (C^{o}_{pq} + C^{(\alpha)}_{pq})\epsilon_q, \tag{4}$$
$$p,q = 1,2 \quad \text{and} \quad 6,$$

where C^{o}_{pq} is the stiffness matrix of the undamaged composite and $C^{(\alpha)}_{pq}$,

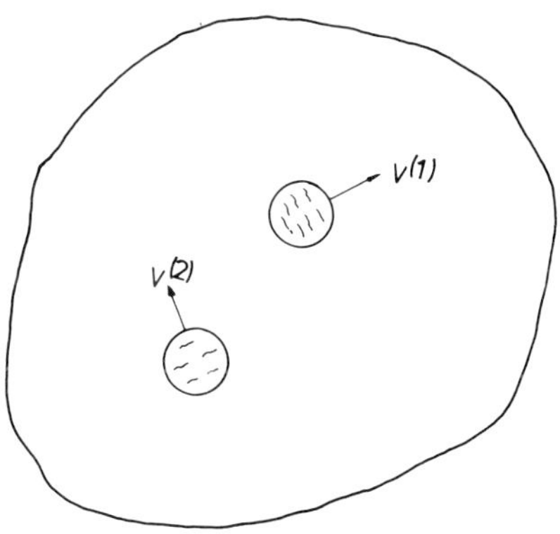

Figure 8. A solid containing two parallel crack arrays.

$\alpha = 1,2, \ldots m$, are the matrices whose coefficients are functions of components of the damage vectors $\mathbf{V}^{(\alpha)}$.

For an initially orthotropic two-dimensional solid containing one array of cracks represented by vector components V_1 and V_2, the stress-strain relations are given by

$$\sigma_p = (C^o_{pq} + C^1_{pq})\epsilon_q \tag{5}$$

where

$$C^o_{pq} = \begin{bmatrix} 2k_1 & k_2 & 0 \\ k_2 & 2k_6 & 0 \\ 0 & 0 & 2k_{10} \end{bmatrix} \tag{6}$$

and

$$C^1_{pq} = \begin{bmatrix} (2k_3V_1^2 + 2k_4V_2^2) & (k_{13}V_1^2 + k_{14}V_2^2) & k_5V_1V_2 \\ (k_{13}V_1^2 + k_{14}V_2^2) & (2k_7V_1^2 + 2k_8V_2^2) & k_9V_1V_2 \\ k_5V_1V_2 & k_9V_1V_2 & (2k_{11}V_1^2 + 2k_{12}V_2^2) \end{bmatrix} \tag{7}$$

where k_i are material constants.

Equations (5–7) show that an initially orthotropic solid will become generally anisotropic as a result of cracking.

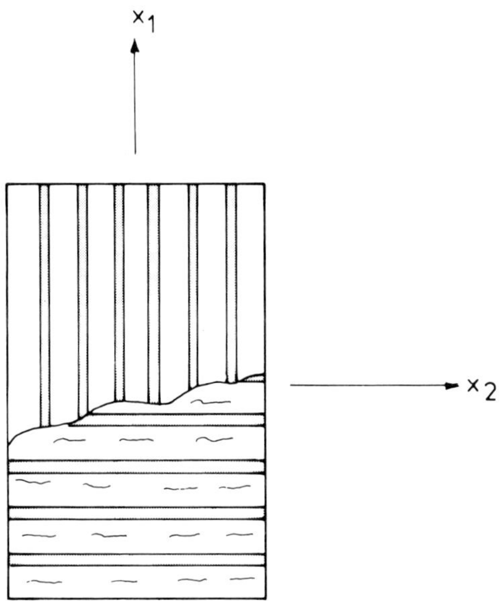

Figure 9. A cross ply laminate with cracks in transverse plies.

Consider for illustration a cross-ply laminate with cracks in the transverse plies (Figure 9). The damage vector for this case is given by

$$\mathbf{V} = D(1,0) \ . \tag{8}$$

The matrix C^1_{pq} becomes

$$C^1_{pq} = \begin{bmatrix} 2k_3D^2 & k_{13}D^2 & 0 \\ k_{13}D^2 & 2k_7D^2 & 0 \\ 0 & 0 & 2k_{11}D^2 \end{bmatrix} \tag{9}$$

The orthotropic symmetry is thus retained in this case. The residual elastic moduli are given by

$$E_1 = E_1^\circ + 2D^2\left[k_3 + k_7(\nu_{12}^\circ)^2 - k_{13}\nu_{12}^\circ\right]$$

$$E_2 = E_2^\circ + 2D^2\left[k_7 + k_3(\nu_{21}^\circ)^2 - k_{13}\nu_{21}^\circ\right]$$

$$\nu_{12} = \nu_{12}^\circ + D^2\left[\frac{1 - \nu_{12}^\circ\nu_{21}^\circ}{E_2^\circ}\right](k_{13} - 2k_7\nu_{12}^\circ)$$

and

$$G_{12} = G_{12}^\circ + 2D^2k_{11} \ . \tag{10}$$

For cracking in off-axis plies that are placed symmetrically about the longitudinal symmetry axis (Figure 10) in an orthotropic laminate, the damage vectors are given by

$$\mathbf{V}^{(1)} = D_1(\sin\theta, \ -\cos\theta) \tag{11}$$

and

$$\mathbf{V}^{(2)} = D_2(\sin\theta, \ \cos\theta), \tag{12}$$

where θ is the off-axis angle.

Expressions for stiffness co-efficients, given in Talreja [10], show that in general the orthotropic symmetry is lost. However, for $\theta = 45$ and $D_1 = D_2$, the orthotropic symmetry is retained. For any other θ the coupling coefficients in the stiffness matrix depend on the damage vector magnitudes. For small magnitudes we may neglect the coupling effects and assume the orthotropy to

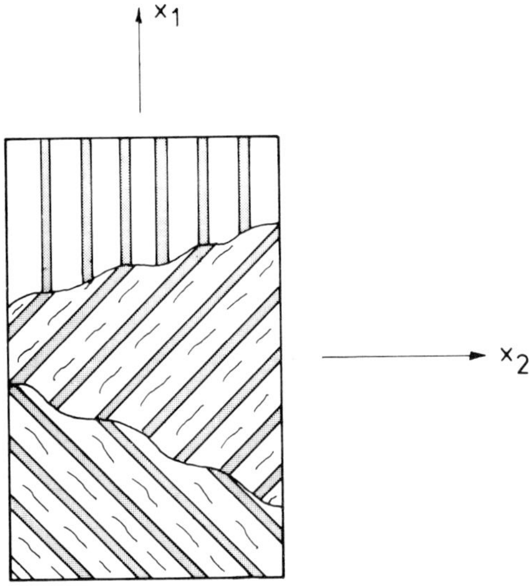

Figure 10. A laminate with cracks in two symmetrically placed off-axis plies.

be maintained. The residual elastic moduli will then be given by Equations (10) when D^2 is replaced by $D_1^2 + D_2^2$.

Elastic Moduli Predictions

It will be fair to state that there exists no scientific method for predicting elastic moduli changes in composite laminates due to transverse cracking. The hitherto used method is commonly known as the ply discount scheme, see for instance Tsai and Hahn [19]. According to this scheme, a cracked off-axis ply is assumed to have lost its transverse and shear stiffnesses and therefore, taking these stiffness co-efficients as zero, the stiffness coefficients of the laminate are recalculated. It is believed that the elastic moduli thus calculated provide limiting values of the residual laminate elastic moduli. Highsmith and Reifsnider [18] found that the elastic moduli calculated by the ply discount scheme gave good approximations to the observed residual elastic moduli of cross-ply laminates of glass/epoxy at saturation cracking (CDS). For $(0, \pm 45)_s$ glass/epoxy laminate, they took only the transverse elastic modulus of the 45° plies to be zero and found the agreement between the observed and predicted residual elastic moduli at CDS not to be good and attributed this to the effect of the free-edge delamination. Highsmith and Reifsnider tried also predicting the longitudinal elastic modulus of a $(0,90_3)_s$ glass/epoxy laminate at various crack densities by reducing the transverse modulus of the 90° plies in proportion to the reduction in the contribution of these plies to carrying the total longitudinal load which they estimated by a shear lag analysis. The residual elastic modulus values calculated by this modification of the ply discount scheme were found to be higher than the observed values at all crack densities.

The stiffness-damage relationships developed by the author provide means for calculating residual elastic moduli of a laminate from crack densities when the elastic moduli of the undamaged laminate and certain material constants are known. For othrotropic laminates the number of materials constants are four (see Equation (10)). Three of these constants, viz. k_3, k_7 and k_{13} can be determined by using the measured values of the two Young's moduli and a Poisson's ratio at undamaged state and at a damage state for which the crack density is known. (The vector magnitude D is given by Equation (2)). The constant k_{11} requires, similarly, the shear moduli and crack density measurements. A laminate configuration which is quite suitable for determining the constants is the cross-ply configuration since only one crack array is involved and a laminate of this configuration remains orthotropic after transverse cracking.

In the following, we shall use the procedure just described to determine the material constants and using these, we shall predict elastic moduli at various crack densities for different laminate configurations.

GLASS/EPOXY LAMINATES

Highsmith and Reifsnider [18] reported crack densities and elastic moduli for various glass/epoxy laminates. Their investigation is the first, and perhaps the only one, that reports changes in all four elastic moduli due to transverse cracking in laminates. We have used the initial moduli and the residual moduli at CDS of $(0,90_3)_s$ laminate along with the saturation crack density to determine the material constants. Due to the uncertainty in measurement of the shear modulus [13] the constant k_{11} is not determined. However, this does not affect determination of the remaining constants, as noted above, and consequently, the predictions of the other elastic moduli are unaffected. The material constants are found to be

$$k_3 = -6.713 \text{ GPa}, \; k_7 = -0.762 \text{ GPa, and}$$
$$k_{13} = -4.467 \text{ GPa}.$$

We note that in determining the damage vector magnitude D, given by Equation (2), we have assumed the average crack length $\bar{\ell}_c$ to be given by the fiber length in the off-axis cracking ply in accordance with the observation that the transverse cracks initiated at the free edges grow soon to span the distance from edge to edge [13]. The thickness of the adjacent plies of same off-axis orientation is taken as the average crack width, $\bar{w}_c$.

We have reported predictions of the longitudinal and transverse Young's moduli and the major Poisson's ratio at CDS for $(90_3,0)_s$, $(0,90)_s$ and

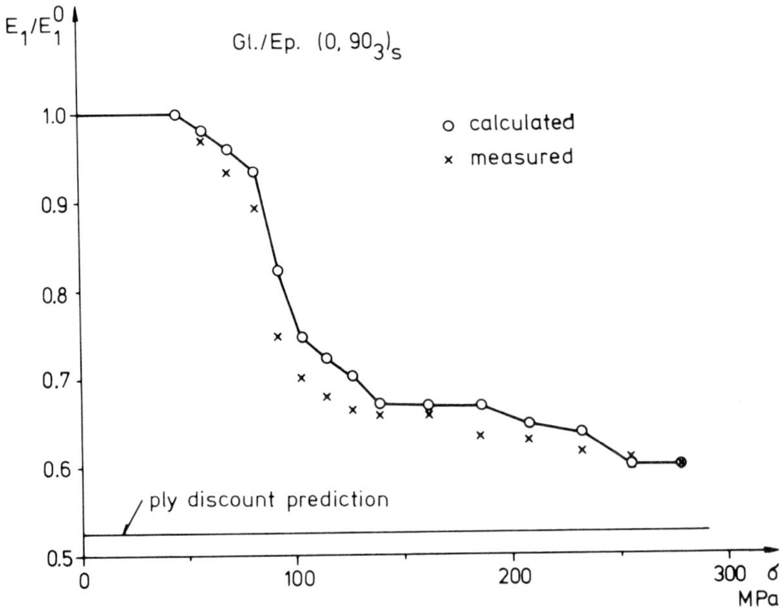

Figure 11. Variation of the longitudinal Young's modulus of $(0,90_3)_s$ laminate with the applied stress.

(0, ± 45)$_s$ laminates previously [11]. We shall here consider the degradation in the longitudinal Young's modulus associated with the development of transverse cracking from initiation to attainment of CDS in the laminates.

Figure 11 shows the degradation of the longitudinal Young's modulus in a $(0,90_3)_s$ laminate with the applied tensile stress. The observed and the predicted values are found to agree fairly well in the entire range of cracking. The prediction by the ply discount scheme is found to slightly overestimate the total modulus reduction. The stress-strain plot, constructed from the modulus reduction data, is shown in Figure 12. The first knee indicating the initiation of transverse cracking appears at a strain of 0.22%. An abrupt increase in strain and the consequent stiffness drop are, however, delayed until 0.45% strain. The subsequent behavior is approximately linear indicating thereby a slow development in the transverse cracking. The behavior displayed here corresponds to the low constraint case, type B, in our classification scheme.

Figure 13 shows the observed and the predicted values of the longitudinal Young's modulus for a $(90_3,0)_s$ laminate along the applied stress axis. The modulus reduction is found to take place earlier than in the $(0,90_3)_s$ laminate probably due to the lack of constraint to the cracking plies from one side.

The longitudinal modulus degradation in fatigue under cyclic tensile stress for a $(0,90)_s$ laminate is shown in Figure 14. The modulus predictions are in

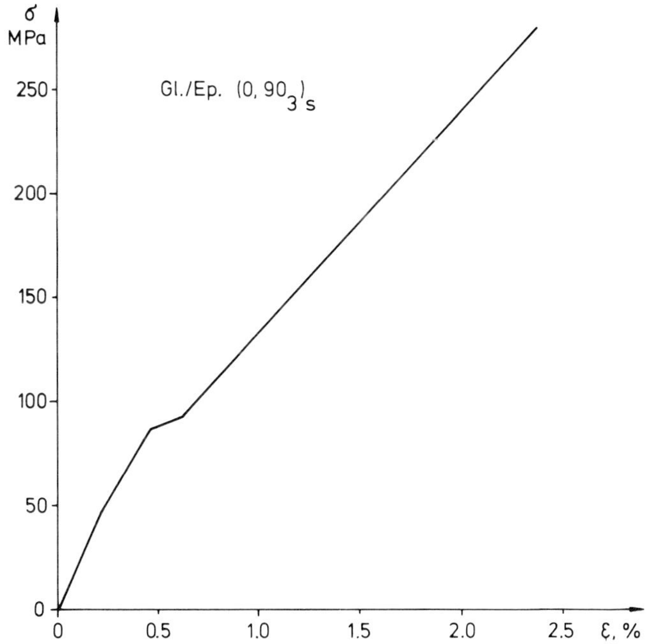

Figure 12. Longitudinal stress-strain diagram for $(0,90_3)_s$ laminate.

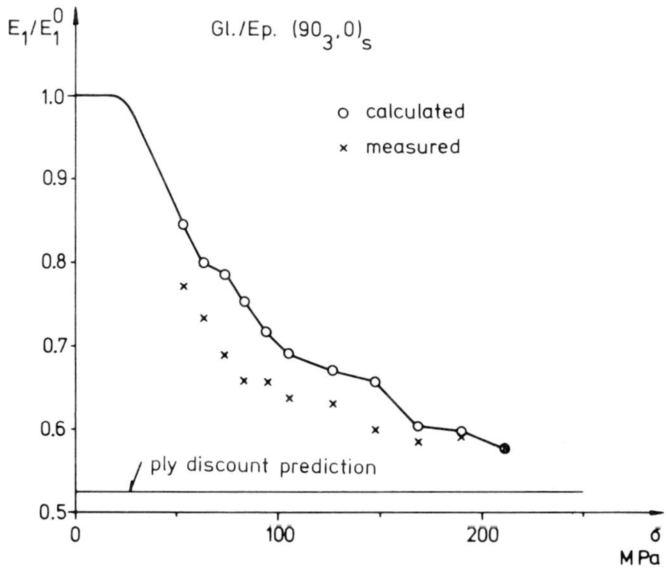

Figure 13. Variation of the longitudinal Young's modulus of $(90_3,0)_s$ laminate with the applied stress.

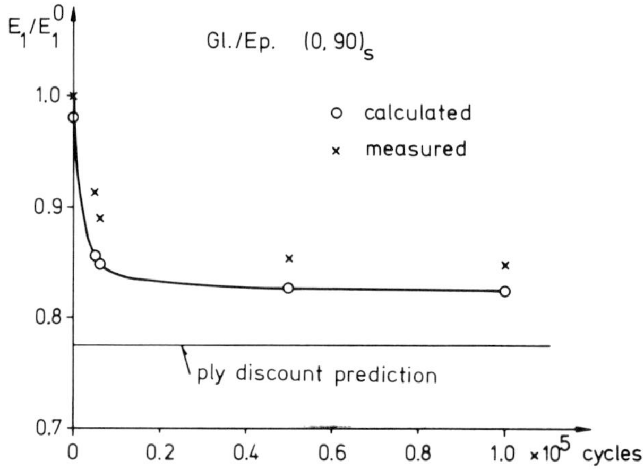

Figure 14. Variation of the longitudinal Young's modulus of $(0,90)_s$ laminate with the applied tension-tension cycles.

reasonable agreement with the observations. The ply discount scheme overestimates the modulus reduction more for this laminate than for the previous two laminates. The stress-strain behavior is displayed in Figure 15. The first knee indicating crack initiation is found to appear at 0.3% strain compared to 0.22% strain for the $(0,90_3)_s$ laminate. Furthermore, no abrupt increase in strain is displayed in contrast to the $(0,90_3)_s$ laminate. The stress-strain behavior of $(0,90)_s$ laminate is thus characteristic of the high constraint case, type C.

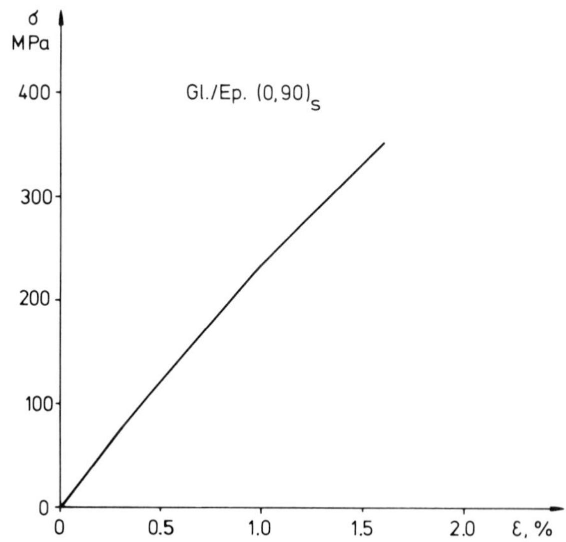

Figure 15. Longitudinal stress-strain diagram for $(0,90)_s$ laminate.

It may be noted at this point that the ply discount scheme gave higher overestimation of the modulus reduction for $(0,90)_s$ laminate than for $(0,90_3)_s$ laminate. This could be associated with the fact that there is higher constraint to transverse cracking in $(0,90)_s$ laminate than in $(0,90_3)_s$ laminate. Thus the cracking plies would be less ineffective in contributing to the laminate stiffness in $(0,90)_s$ laminate than in $(0,90_3)_s$ laminate.

The last laminate of glass/epoxy considered is the $(0, \pm 45)_s$ laminate for which the predicted and the observed modulus reductions under tension-tension fatigue are shown in Figure 16. The predictions by our method are in good agreement with the observations. The ply discount scheme used in the usual manner, i.e. with the transverse and the shear moduli of the 45° plies set to zero, gives a considerable overestimation of the total modulus reduction. However, the modified ply discount scheme as used by Highsmith and Reifsnider [18], i.e. with only the transverse modulus of the 45° plies set to zero, underestimates the total modulus reduction. The large difference between predictions by the two versions of the ply discount scheme show the uncertainties involved in application of this scheme.

GRAPHITE/EPOXY LAMINATES

Kistner, Whitney and Browning [20] have recently reported crack densities and stress-strain diagrams for various graphite/epoxy laminates. We shall predict the elastic moduli for their laminates using our method and compare our predictions with the observations.

As discussed above, we must determine three material constants in order to

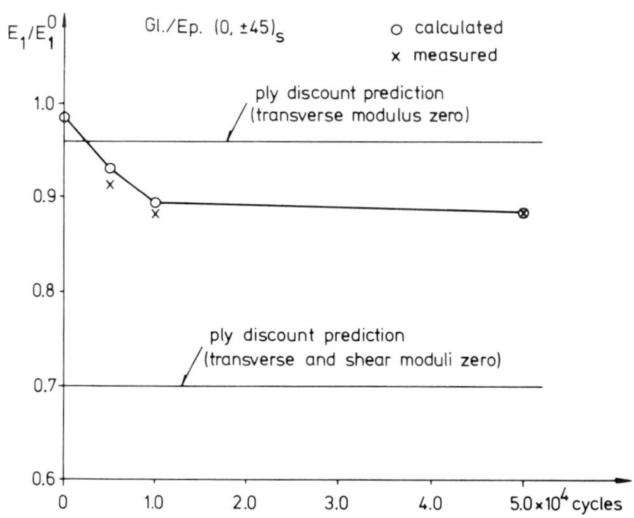

Figure 16. Variation of the longitudinal Young's modulus of $(0, \pm 45)_s$ laminate with the applied tension-tension cycles.

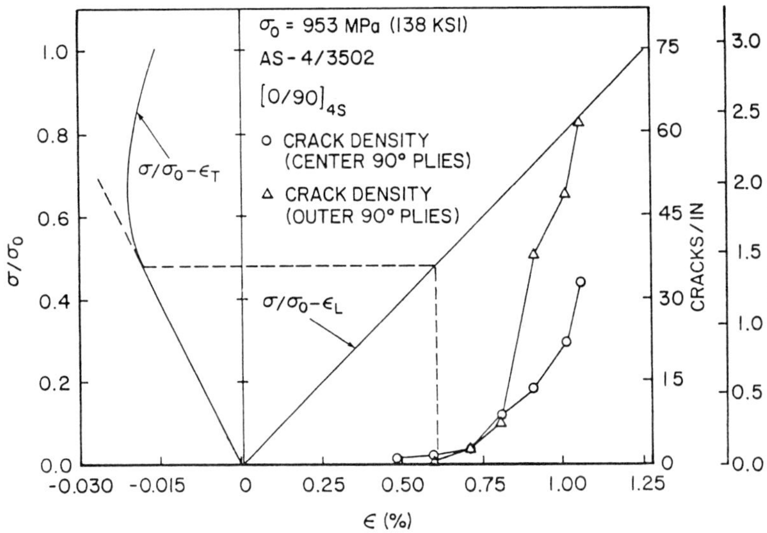

Figure 17. Stress-strain diagram and crack densities for $(0,90)_{4S}$ laminate. (From [20]).

predict Young's moduli and Poisson's ratios for given crack densities. We have used the data reported by Kistner, et al. for $(0,90)_{4s}$ laminate to determine the constants. The stress-strain diagrams for this laminate are shown in Figure 17 along with the corresponding crack densities. We have used the longitudinal Young's modulus and the Poisson's ratio at the maximum crack density and have assumed the transverse Young's modulus to remain unchanged. The constants thus determined are

$$k_3 = -17.857 \text{ GPa}, \; k_7 = -0.141 \text{ GPa}$$

and

$$k_{13} = -5.557 \text{ GPa}.$$

Using these constants, the Poisson's ratio values at other crack densities are calculated and shown in Figure 18 and compared with the measured values along the applied stress axis. The predictions of the maximum reduction in the Poisson's ratio by the ply discount scheme is also shown. It is of interest here to notice that large changes, up to 40 percent, are found in the Poisson's ratio, although the longitudinal Young's modulus does not change much. This serves to illustrate the necessity of considering all moduli changes in order to characterize the elastic response of cracked laminates.

The stress-strain diagram and the crack densities for a $(45,90,-45,90,45,90\ -45,\overline{90})_s$ laminate, as reported by Kistner et al., are shown in Figure 19. The behavior displayed here is characteristic of the low constraint cracking, type B. The development of transverse cracking is accompained by a large increase in strain which seems to slow down as the saturation state is approached. The predicted values of the longitudinal Young's modulus are compared with the experimental values as measured from the stress-strain plot (Figure 19) and are shown plotted against the applied stress in Figure 20. The prediction of the lowest residual modulus by the ply discount scheme is also shown here. It is of interest to notice that this prediction does not account for the maximum loss of modulus in this case.

Figure 21 shows the stress-stress plot and the crack densities for $(60,90-60,90,60,90,-60,\bar{90})_s$ laminate. The first cracking is accompanied by an abrupt increase in strain and is followed by a gradual increase in strain as more cracking occurs. The knee in the stress-strain curve occurs at 0.55% strain which is approximately the failure strain for the unconstrained 90° plies. The constraint to transverse cracking by the 60° plies is thus very low and may be characterized as approaching the limiting case of no constraint, type A. The predicted and the experimental values of the longitudinal Young's modulus are shown plotted against the applied stress in Figure 22. The predictions lie above the observed values of the modulus beyond the first cracking, as expected, since the crack densities used in calculations are of 90° plies only while cracking in the 60° plies was also observed [20]. The ply discount prediction of the minimum value of the modulus does not account for the total loss of modulus due to cracking in the 90° plies.

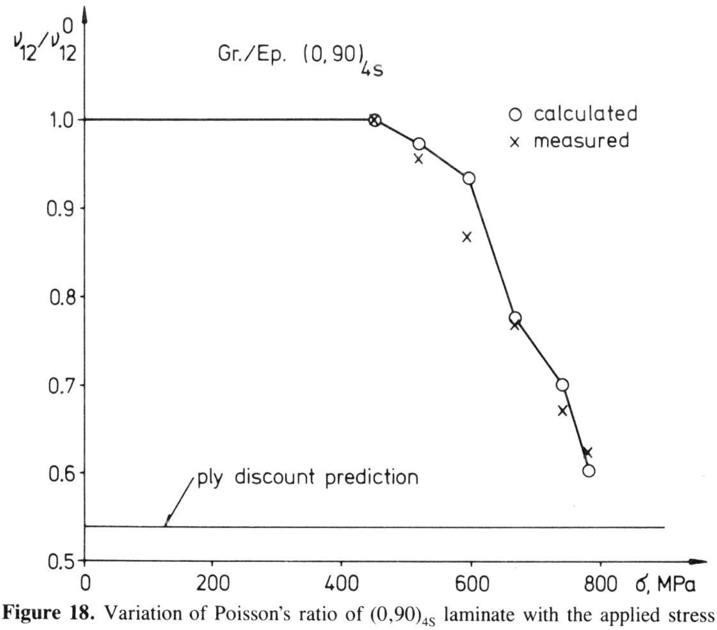

Figure 18. Variation of Poisson's ratio of $(0,90)_{4S}$ laminate with the applied stress.

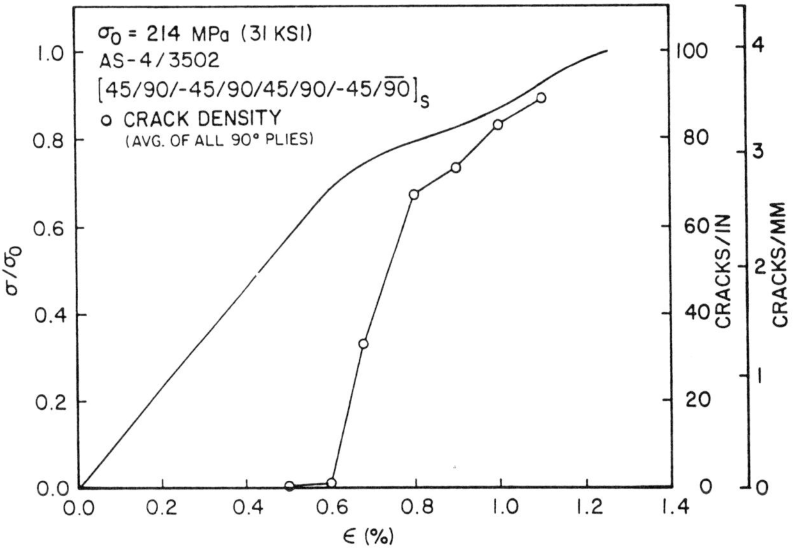

Figure 19. Stress-strain diagram and crack densities for $(45,90,-45,90,45,90,-45,\overline{90})_s$ laminate. (From [20]).

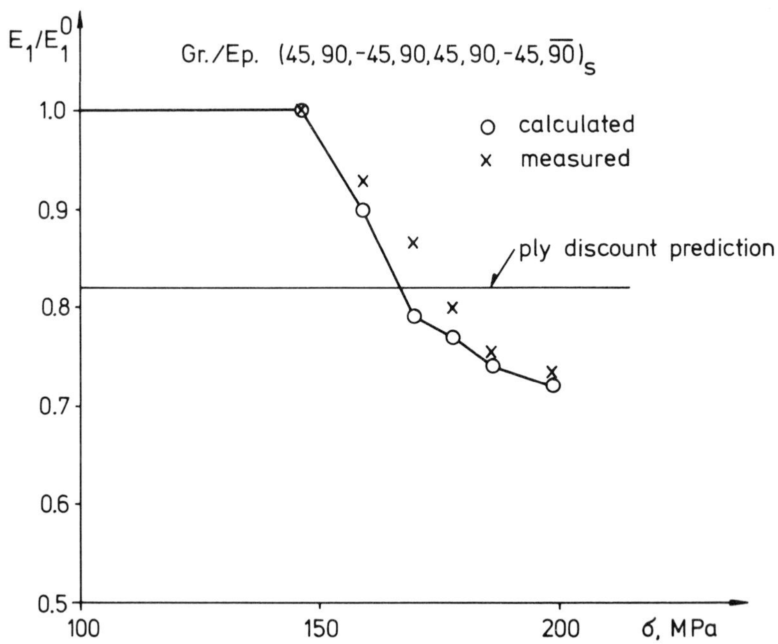

Figure 20. Variation of the longitudinal Young's modulus of $(45,90,-45,90,45,90-45,90)_s$ laminate.

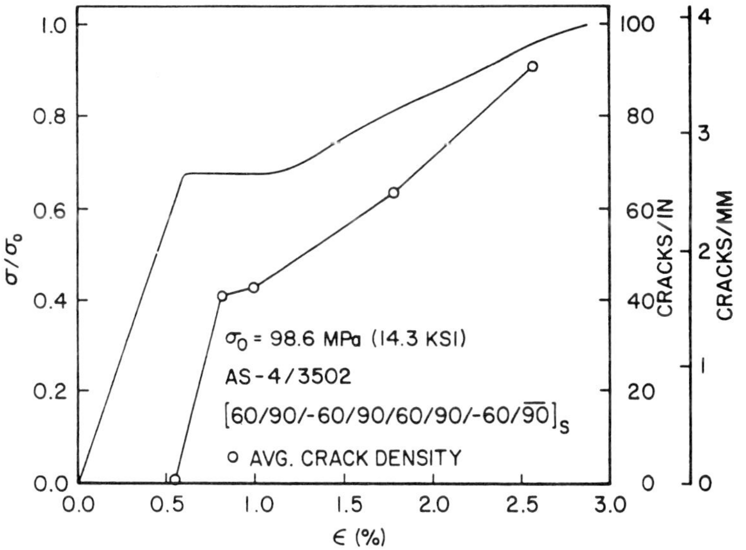

Figure 21. Stress-strain diagram and crack densities for $(60,90-60,90,60,90,-60,\bar{9}0)_s$ laminate. (From [20]).

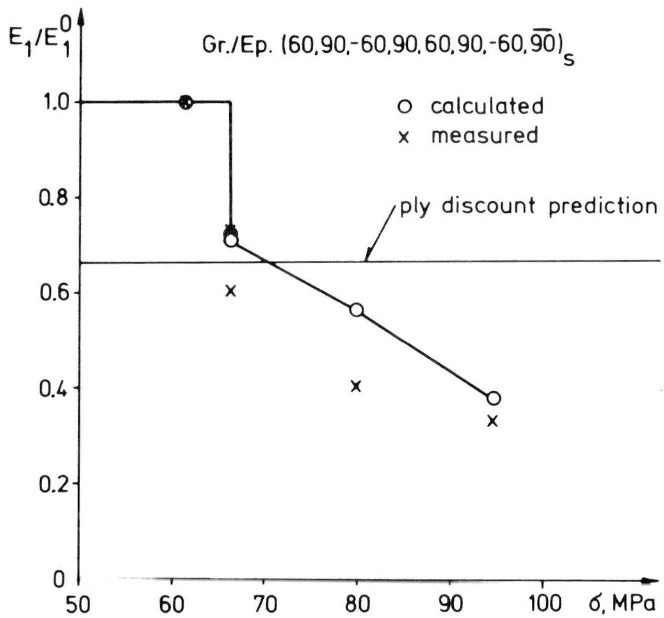

Figure 22. Variation of the longitudinal Young's modulus of $(60,90,-60,90,60,90,-60,\bar{9}0)_s$ laminate.

Conclusions

The classification of the transverse cracking effects in the four types proposed here is intended as a useful means for a qualitative assessment of the laminate performance subsequent to cracking.

The method described here for quantitative predictions of the laminate stiffness changes due to transverse cracking has been demonstrated to have good accuracy. The material constants which must be determined in order to apply this method are easily calculated for the case of unchanged orthotropic symmetry by using elastic moduli measurements for a given crack density. The cross-ply laminate configuration is a good choice for determining the constants, since this laminate retains its orthotropic symmetry after cracking, has only one array of cracks and suffers negligible reduction in the transverse Young's modulus.

In the case of highly constrained transverse cracking, as illustrated by the $(0,90)_{4s}$ graphite/epoxy laminate, the Poisson's ratio may change significantly, although little change may occur in the Young's moduli. This serves to emphasize the necessity of considering all moduli changes in order to characterize the elastic response of laminates undergoing transverse cracking. The method presented here is a useful tool for achieving this.

The ply discount scheme has been demonstrated to be unreliable for predicting minimum values of moduli for cracked laminates. From the cases considered here, it appears that the ply discount predictions overestimate moduli reductions for high constraint cracking and underestimate these for low constraint cracking. For full validity of this trend, further investigation seems necessary.

References

1. Broutman, L. J. and Sahu, S., "Progressive Damage of a Glass Reinforced Plastic During Fatigue," *SPI, 24th Annual Technical Conference*, Washington, D.C., Section 11-D (February 1969).
2. Garrett, K. W. and Bailey, J. E., "Multiple Transverse Fracture in 90° Cross-Ply Laminates of a Glass Fibre-Reinforced Polyester," *Journal of Materials Science*, 12, pp. 157–168 (1977).
3. Aveston, J. and Kelly, A., "Theory of Multiple Fracture of Fibrous Composites," *Journal of Materials Science*, 8, pp. 352–362 (1973).
4. Parvizi, A., Garrett, K. W. and Bailey, J. E., "Constrained Cracking in Glass Fibre-Reinforced Epoxy Cross-Ply Laminates," *Journal of Materials Science*, 13, pp. 195–201 (1978).
5. Bader, M. G., Bailey, J. E., Curtis, P. T. and Parvizi, A., "The Mechanisms of Initiation and Development of Damage in Multi-Axial Fibre-Reinforced Plastics Laminates," *Proceedings of the Third International Conference on Mechanical Behavior of Materials*, Cambridge, England, Vol. 3, pp. 227–239 (August 1979).
6. Wang, A. S. D. and Crossman, F. W., "Initiation and Growth of Transverse Cracks and Edge

Delamination in Composite Laminates, Part 1. An Energy Method," *Journal of Composite Materials*, Supplement, 14, pp. 71–87 (1980).

7. Crossman, F. W., Warren, W. J., Wang, A. S. D. and Law, G. E., Jr., "Initiation and Growth of Transverse Cracks and Edge Delamination in Composite Laminates, Part 2. Experimental Correlation," *Journal of Composite Materials*, Supplement, 14, pp. 88–108 (1980).
8. Reifsnider, K. L., Henneke, E. G. and Stinchcomb, W. W., "Defect-Property Relationships in Composite Materials," *AFML-TR-76-81, Part IV*. Air Force Materials Laboratory (June 1979).
9. Reifsnider, K. L., "Some Fundamental Aspects of Fatigue and Fracture Response of Composite Materials," *Proceeding, 14th Annual Meeting of Society of Engineering Science*, Lehigh University, Bethlehem, PA (November 1977).
10. Talreja, R., "A Continuum Mechanics Characterization of Damage in Composite Materials," to appear in *Proc. Royal Soc. London*, Chapter 6 in the present volume.
11. Talreja, R., "Residual Stiffness Properties of Cracked Composite Laminates," *Proc. Sixth Int. Conf. on Fracture*, ICF6, New Delhi, India, Chapter 7 in the present volume (December 4–10, 1984).
12. Talug, A., "Analysis of Stress Fields in Composite Laminated with Interior Cracks," Ph.D. Thesis, College of Engineering, Virginia Polytechnic Institute and State University, Blacksburg, Virginia (September 1978).
13. Highsmith, A. L., "Stiffness Reduction Resulting from Transverse Cracking in Fiber-Reinforced Composite Laminates," M.S. Thesis, College of Engineering, Virginia Polytechnic Institute and State University, Blacksburg, Virginia (August 1981).
14. Aveston, J. and Kelly, A., "Tensile First Cracking Strain and Strength of Hybrid Composites and Laminates," Phil., *Trans. Royal Soc.*, London, Vol. A294, pp. 519–534 (1980).
15. Hahn, H. T. and Tsai, S. W., "On the Behavior of Composite Laminates after Initial Failures," *Journal of Composite Materials*, 8, pp. 288–305 (1974).
16. Gottesman, T., Hashin, Z. and Brull, M. A., "Effective Elastic Moduli of Cracked Fiber Composites," in *Advances in Composite Materials, Proc. Third Int. Conf. on Composite Materials*, Paris, France, A. R. Bunsell, et. al., eds., Pergamon Press, Oxford, pp. 749–758 (1980).
17. Laws, N., Dvorak, G. J. and Hejazi, M., "Stiffness Changes in Unidirectional Composites Caused by Crack Systems," *Mechanics of Materials*, 2, pp. 123–137 (1983).
18. Highsmith, A. L. and Reifsnider, K. L., "Stiffness-Reduction Mechanisms in Composite Laminates," in *Damage in Composite Materials, ASTM STP 775*, American Society for Testing and Materials, Philadelphia, PA, pp. 103–117 (1982).
19. Tsai, S. W. and Hahn, H. T., "Composite Materials Workbook," *Technical Report AFML-TR-78-33*, Air Force Materials Laboratory, Wright-Patterson Air Force Base, Ohio, p. 195 (March 1978).
20. Kistner, M. D., Whitney, J. M. and Browning, C. E., "First Ply Failure of Graphite/Epoxy Laminates," *Proc. U.S.-Japan Conference on Composite Materials*, NASA-Langley, Hampton, Virginia (June 1983).

Part 3

FATIGUE RELIABILITY

CHAPTER 9

Estimation of Weibull Parameters for Composite Material Strength and Fatigue Life Data

Abstract: Two classical methods for estimation of Weibull parameters, namely, the moment estimation and the maximum likelihood estimation are reviewed. It is demonstrated that, assuming the location parameter to be zero, the errors made in estimating both the scale and the shape parameters by these methods increase linearly with the true value of the location parameter.

An alternative method for Weibull parameter estimation is proposed. This method provides estimates for all three parameters when the shape parameter is close to one, and estimates the scale and the shape parameters more accurately than the other two methods for higher values of the shape parameter. Furthermore, unlike the other methods, this method is capable of separating the constituent components in a multi-component sample and estimating the parameters of each component. This property makes the method suitable for statistical analysis of composite material strength and fatigue life data.

Key words: composite materials, Weibull distribution, statistical analysis, strength data, fatigue life data, reliability assessment, fatigue (materials).

IN ASSESSING THE reliability of composite structures, Weibull distribution has proved to be a useful and versatile means of describing composite material strength and fatigue life data. However, proper use of this distribution requires that its parameters be accurately estimated. Furthermore, fracture of composite materials is complex, which often results in multi-component strength and fatigue life data. This necessitates using an estimation procedure that is capable of detecting heterogeneity in given data and estimating parameters of each constituent component. The classical methods of parameter estimation, for example, the moment estimation and the maximum likelihood estimation, are incapable of this task. Both these methods require homogeneous samples and are highly inefficient if all three Weibull parameters are to be estimated.

In statistical analyses of composite material strength and fatigue life data

Reprinted by permission of the American Society for Testing and Materials from *Fatigue of Fibrous Composite Materials, ASTM STP 723* (1981).

reported in the literature, a two-parameter Weibull distribution has often been assumed, implying that the location parameter is zero. The consequences of this assumption have not yet been examined. It is therefore demonstrated here that if given data, which actually belong to a three-parameter Weibull population, are assumed to be two-parameter Weibull data, then the estimates of these two parameters from these data will be higher than their true values. Furthermore, the errors made in so estimating the two parameters will increase linearly with the true value of the location parameter.

In the following, we shall describe a method that allows estimation of all three parameters except when the estimate of the location parameter becomes negative. In the latter case, requiring the location parameter to be zero or positive, good estimates of the other two parameters are obtained. Furthermore, with this method, the constituent components in heterogeneous data can be separated and parameters for each component can then be estimated.

The superiority of the proposed method is demonstrated by comparing its estimates from random samples with the estimates of the moment estimation and the maximum likelihood methods from the same samples. This is done for two values of the shape parameter using five samples of ten data points.

Finally, composite material strength and fatigue life data are taken from the literature to illustrate the use of the proposed method.

Estimation of Weibull Parameters

The Weibull distribution function is given by

$$F(x\colon a, b, c) = 1 - \exp\left\{ -\left(\frac{x - a}{b} \right)^c \right\} \tag{1}$$

for $x > a$, $b > 0$, and $c > 0$, where a, b, and c are the location, scale, and shape parameters, respectively.

For strength distributions, $a < 0$ will not be acceptable for tensile strength. For fatigue life distributions, $c < 1.0$ will not be acceptable, as that would lead to a decreasing failure rate.

A set of observations, for example, test data, is called a sample and, when arranged in increasing order, an individual observation is called an order statistic denoted by x_i, such athat $x_1 < x_2 < x_3 \ldots < x_N$. A complete statistical analysis of a given sample consists of the following steps [1]:

1. testing the homogeneity of the sample,
2. testing the acceptability of one or more of the assumed distributions,
3. selecting one out of a set of acceptable distributions, and
4. estimating the unknown parameters of the accepted distribution.

Herein we shall discuss Step 4 with Step 1 forming part of one of the methods of estimation discussed.

MOMENT ESTIMATION

This is a classical method introduced by K. Pearson in 1894. Essentially, it consists of equating as many of the population moments to the sample moments as the number of unknown parameters.

The sample moments are given by

$$A_k = \frac{\Sigma(x_i)^k}{N} \tag{2}$$

and

$$B_k = \frac{\Sigma(x_i - \bar{x})^k}{N} \tag{3}$$

where

$$\bar{x} = \frac{\Sigma x_i}{N} \tag{4}$$

and A_k are the moments about origin and B_k are the central moments.

The population moments are given by

$$\alpha_k = \int_{-\infty}^{\infty} x^k f(x) dx \tag{5}$$

and

$$\beta_k = \int_{-\infty}^{\infty} (x - \bar{x})^k f(x) dx \tag{6}$$

where $f(x)$ is the Weibull density function given by

$$f(x) = \frac{c}{b}\left(\frac{x - a}{b}\right)^{c-1} \exp\left\{-\left(\frac{x - a}{b}\right)^c\right\} \tag{7}$$

The first three population moments are thus given by

$$\alpha_1 = a + bg_1 \tag{8}$$

$$\alpha_2 = a^2 + b^2 g_2 + 2abg_1 \tag{9}$$

$$\alpha_3 = a^2 + b^3 g_3 + 3abg_2 + 3a^2 bg_1 \tag{10}$$

$$\beta_1 = 0 \tag{11}$$

$$\beta_2 = b^2(g_2 - g_1^2) \tag{12}$$

$$\beta_3 = b^3(g_3 - 3g_1g_2 + 2g_1^3) \tag{13}$$

where

$$g_i = \left(\frac{i}{c}\right)! \tag{14}$$

For a two-parameter Weibull distribution, the parameters b and c may be estimated as follows.

Equating A_1 to α_1 and taking $a = 0$, we have

$$bg_1 = \bar{x} = \frac{\Sigma x_i}{N} \tag{15}$$

Equation (15) gives the estimate of b as

$$b^* = \frac{\Sigma x_i}{N} \tag{16}$$

To obtain an estimator for c, we form the ratio

$$V = \frac{(B_2)^{1/2}}{A_1} = \frac{[\Sigma(x_i - \bar{x})^2]^{1/2}}{\sqrt{N} \cdot \bar{x}} \tag{17}$$

The corresponding ratio for population moments is given by

$$K = \frac{(\beta_2)^{1/2}}{\alpha_1} = \frac{(g_2 - g_1^2)^{1/2}}{g_1} \tag{18}$$

which is only a function of c, according to Equation (14). Thus, the value of c satisfying

$$K = V \tag{19}$$

is taken as an estimate of c.

The values of K have been tabulated by Weibull and Weibull [1] for different values of c.

For three-parameter Weibull distribution, the estimators are obtained as follows.

The coefficients of skewness for sample and population are given by

$$G_s = \frac{B_3}{(B_2)^{3/2}} = \sqrt{N}\,\frac{\Sigma(x_i - \bar{x})^3}{[\Sigma(x_i - \bar{x})^2]^{3/2}} \tag{20}$$

and

$$G_p = \frac{\beta_3}{(\beta_2)^{3/2}} = \frac{g_3 - 3g_1g_2 + 2g_1{}^3}{(g_2 - g_1{}^2)^{3/2}} \tag{21}$$

The estimate of c is thus the value that satisfies the equality

$$G_s = G_p \tag{22}$$

The values of G_p have been tabulated by Gumbel [2] and by Weibull and Weibull [1] for different values of c.

Equating β_2 to B_2, we have

$$b^2(g_2 - g_1{}^2) = \frac{\Sigma(x_i - \bar{x})^2}{N} \tag{23}$$

that gives an estimate of b as

$$b^* = \frac{[\Sigma(x_i - \bar{x})^2]^{1/2}}{N^{1/2}(g_2 - g_1{}^2)^{1/2}} \tag{24}$$

in which g_i are calculated from the estimates of c.

The estimate of a is given by equating α_1 to A_1. Thus

$$a^* = \bar{x} - b^*\left(\frac{1}{c^*}\right)! \tag{25}$$

MAXIMUM LIKELIHOOD ESTIMATION

The likelihood function $L(a, b, c)$ for the first M order statistics from a sample of size N is given by

$$L_m = lnL(a,b,c) = lnN! - ln(N - M)! + M(ln\,c - c\,lnb)$$

$$+ (c - 1)\sum_{i=1}^{M} ln(x_i - a) - \sum_{i=1}^{M} [(x_i - a)/b]^c$$

$$- (N - M)[(x_N - a)/b]^c \tag{26}$$

The maximum likelihood equations are obtained by equating to zero the partial derivatives of *ln L* with respect to each of the three parameters. Thus

$$\frac{\partial L_m}{\partial a} = cb^{-c}\left[\sum_{i=1}^{M} (x_i - a)^{c-1} + (N - M)(x_M - a)^{c-1}\right]$$

$$- (c - 1)\sum_{i=1}^{M} (x_i - a)^{-1} = 0 \tag{27}$$

$$\frac{\partial L_m}{\partial b} = cb^{-(c+1)}\left[\sum_{i=1}^{M} (x_i - a)^{c} + (N - M)(x_M - a)^{c}\right]$$

$$- Mcb^{-1} = 0 \tag{28}$$

$$\frac{\partial L_m}{\partial c} = \frac{M}{c} + \sum_{i=1}^{M} ln\left[\frac{(x_i - a)}{b}\right] - \sum_{i=1}^{M}\left[\left(\frac{x_i - a}{b}\right)^{c} ln\left(\frac{x_i - a}{b}\right)\right]$$

$$- (N - M)\left[\frac{x_M - a}{b}\right]^{c} ln\left[\frac{x_m - a}{b}\right] = 0 \tag{29}$$

Antle and Klimko [3] have discussed solutions of the likelihood equations. They have shown that if a solution (a^*, b^*, c^*) is a local maximum, then another solution (a', b', c') exists that corresponds to a saddle point. They have also demonstrated by a numerical example that even if a solution is a local maximum, it may not be the maximum likelihood estimate.

Harter and Moore [4] have developed an iterative procedure that involves estimating the three parameters, one at a time, in the cyclic order *b, c, a*. They claim that positive values of b^* and c^* can always be found in this way. In estimating a, however, they find that no value of a in the permissible interval $0 \leq a \leq x_1$ will satisfy the likelihood Equation (29). By censoring x_1 in such a case, they succeed in finding the estimates.

Antle and Klimko [3], however, do not recommend maximum likelihood estimation of all three parameters for sample size $\leq$ 10. Weibull and Weibull [1] used a computer program provided by Antle for maximum likelihood estimation of 300 random samples of size 10 and 20 each drawn from Weibull populations (0,1,1), (0,1,2), and (0,1,5). They found that for sample size 10 and for $c = 2$ and $c = 5$, about half the estimates were unacceptable, that is, they were not the local maximum solutions.

For a two-parameter ($a = 0$) complete sample ($N = M$), the likelihood function on eliminating b from Equations (27–29) becomes

$$L_m(c) = lnc - ln\Sigma x_i^c + (1/N)[(c - 1)\Sigma lnx_i] \qquad (30)$$

The maximum likelihood equation for estimating c is then given by

$$\frac{1}{c} - \frac{\Sigma x_i^c lnx_i}{\Sigma x_i^c} + \frac{\Sigma lnx_i}{N} = 0 \qquad (31)$$

The maximum likelihood estimate of b is given by Equation (28) as

$$b^* = \left[\frac{\Sigma x_i^{c*}}{N}\right]^{1/c*} \qquad (32)$$

For maximum likelihood estimation of two-parameter Weibull distribution, Cohen [5] has given variance-covariance matrices for censored and complete samples and Thoman, Bain, and Antle [6] and McCool [7] have given confidence intervals for estimates and unbiasing factors for shape parameter estimates.

STANDARDIZED VARIABLE ESTIMATION

The standardized variable z is defined by

$$z = \frac{x - a}{b} \qquad (33)$$

The distribution function of the standardized variable is given by $F(z;\ 0,\ 1,\ c)$. Thus the order statistics, z_i, are independent of a and b, and depend only on the shape parameter, c.

The expected value, Ez_i, the median, Mz_i, and the variance, Vz_i, of the order statistics, z_i, were derived by Lieblein [8], and are given by

$$Ez_i = iC_i^N\left(\frac{1}{c}\right)!\sum_{j=0}^{i-1}(-1)^{i-1-j}C_j^{i-1}(N-j)^{-(1+1/c)} \qquad (34)$$

$$Mz_i = [-ln(1 - MP_i)]^{1/c} \qquad (35)$$

$$Vz_i = E(z_i^2) - [E(z_i)]^2 \qquad (36)$$

where

$$E(z_i^2) = iC_i^N\left(\frac{2}{c}\right)!\sum_{j=0}^{i-1}(-1)^{i-1-j}C_j^{i-1}(N-j)^{-(1+2/c)} \qquad (37)$$

C_i^j are the binomial coefficients and MP_i are the median percentage points.

As seen in Equations (34–37), all the characteristic values of the order statistics, z_i, depend only on the sample size, N, and the shape parameter, c. If, now, the order statistics, z_i, are estimated by their expected values, Ez_i, or by their median, Mz_i, we obtain from Equation (33)

$$x_i = a + bEz_i \tag{38}$$

or

$$x_i = a + bMz_i \tag{39}$$

Thus, if the observations, x_i, are plotted against Ez_i or Mz_i for the true value of c, the data points (x_i, Ez_i) or (x_i, Mz_i) will, with due regard to the sampling scatter, fall along a straight line. The estimate of c may then be taken as the value for which the linear regression gives the maximum correlation coefficient. Once c has been estimated, the estimates of a and b will be given by the x_i-intercept and the slope, respectively, of the best-fit line.

The tables for Ez_i and Vz_i are given by Weibull [9] and for Mz_i by Weibull [10] for different values of c and samples sizes.

It was shown by Weibull and Weibull [10] that a good approximation to MP_i is given by the simple formula

$$MP_i = k(i)/[k(i) + k(N + 1 - i)] \tag{40}$$

where

$$k(i) = i - 0.334 + 0.0252/i \tag{41}$$

The computation of Mz_i is thus considerably simplified and preference may therefore be given to Equation (39) over Equation (38) for estimation of parameters.

It may be noted that this method of estimating parameters can be used also for censored or truncated samples.

The linear relationships in Equations (38) and (39) indicate that if the observations, x_i, belong to different populations, then in the (x_i, Ez_i) or (x_i, Mz_i) plots the data points will scatter about different straight lines. An inspection of these plots would then allow us to separate the different components in the sample and estimate parameters for each component by finding the shape parameter values that maximize correlation coefficients for the corresponding straight lines. This will be illustrated by analyzing actual composite strength data in the sequel.

Table 1. Random samples drawn from a Weibull distribution (0, 1, 1).

i	x_i	x_i	x_i	x_i	x_i
1	0.315	0.013	0.200	0.024	0.027
2	0.342	0.229	0.236	0.272	0.221
3	0.589	0.233	0.333	0.574	0.265
4	0.590	0.244	0.401	0.772	0.277
5	1.117	0.365	0.911	0.920	0.419
6	1.470	0.489	0.966	0.989	0.665
7	1.693	0.606	1.219	1.157	0.980
8	2.041	0.909	1.773	2.264	1.390
9	2.679	1.413	1.789	3.040	1.509
10	3.667	1.773	1.997	3.299	2.647

Estimation from Random Samples

Five random samples of each size 10 were constructed for a chosen Weibull population using the relationship

$$x_i = a + b(-\ln r_i)^{1/c} \tag{42}$$

where r_i are random numbers drawn from a uniform distribution in the interval (0, 1). The order statistics, x_i, are shown in Table 1 for a Weibull population (0, 1, 1). Weibull parameters were estimated from these samples using each of the methods just described.

MOMENT ESTIMATION

All three Weibull parameters were estimated for two populations (0, 1, 1) and (0, 1, 10) using Equations (22), (24), and (25). The estimates are shown in Table 2. As seen here, negative estimates of location parameters are obtained for all samples of the first population and for two samples of the second population. These estimates can therefore not be accepted. The average errors

Table 2. Estimates of Weibull parameters by moment estimation.

Population	(0, 1, 1)			(0, 1, 10)		
Sample	a*	b*/b	c*/c	a*	b*/b	c*/c
1	−0.29	2.07	1.81	0.72	0.32	0.39
2	−0.10	0.92	1.57	−1.81	2.97	3.23
3	−0.60	1.89	2.76	0.60	0.38	0.49
4	−0.52	2.21	1.88	−8.46	9.70	10.0
5	−0.23	1.25	1.49	0.12	0.86	0.87
Average	−0.35	1.67	1.90	−1.77	2.85	3.00
Variance	0..034	0.246	0.205	12.029	12.680	13.34

Table 3. Estimates of Weibull parameters by moment estimation assuming a = 0.

Population	(0.1, 1, 1)		(0.5, 1, 1)		(1.0, 1, 1)		(2.0, 1, 1)	
Sample	b*/b	c*/c	b*/b	c*/c	b*/b	c*/c	b*/b	c*/c
1	1.72	1.51	2.20	1.94	2.76	2.51	3.82	3.66
2	0.80	1.36	1.27	2.20	1.81	3.31	2.84	5.62
3	1.21	1.70	1.67	2.41	2.21	3.33	3.24	5.24
4	1.57	1.35	2.07	1.76	2.64	2.30	3.71	3.42
5	1.00	1.23	1.51	1.80	2.07	2.56	3.13	4.15
Average	1.26	1.43	1.74	2.02	2.30	2.80	3.35	4.42
Variance	0.119	0.026	0.119	0.060	0.125	0.186	0.136	0.747

made in estimating the scale and the shape parameters are higher for the higher value of the shape parameter. The variance of the estimates, the inverse of which defines the efficiency of the estimation method, is higher for the higher value of the shape parameter. It therefore appears that both the accuracy and the efficiency of this estimation method decrease with increasing values of the shape parameter.

Tables 3 and 4 illustrate the errors made in estimating the scale and the shape parameters when the location parameter is taken to be zero. The estimates in these tables are found by using Equations (16) and (19) for Weibull populations (a, 1, 1) and (a, 1, 10) with a having values 0.1, 0.5, 1.0, and 2.0 for each set of five samples. As seen here, the errors made in estimating the scale and the shape parameters increase with increasing value of the location parameter. The average ratios of the estimates to the true parameter values are plotted against the true location parameter values in Figure 1 for $c = 1$ and in Figure 2 for $c = 2$. As shown by these figures, the errors appear to increase linearly with the true location parameter value. Comparing these two figures, it can also be seen that the errors in estimating b and c are lower for higher

Table 4. Estimates of Weibull parameters by moment estimation assuming a = 0.

Population	(0.1, 1, 10)		(0.5, 1, 10)		(1.0, 1, 10)		(2.0, 1, 10)	
Sample	b*/b	c*/c	b*/b	c*/c	b*/b	c*/c	b*/b	c*/c
1	1.15	1.67	1.55	2.27	2.05	3.08	3.04	4.76
2	1.06	1.11	1.45	1.67	1.96	2.16	2.96	3.33
3	1.11	1.67	1.50	2.50	2.01	3.12	3.00	4.76
4	1.14	1.07	1.54	1.43	2.04	2.00	3.04	3.03
5	1.08	1.13	1.48	1.67	1.98	2.17	2.99	3.33
Average	1.11	1.33	1.50	1.91	2.01	2.51	3.01	3.84
Variance	0.001	0.077	0.001	0.165	0.001	0.239	0.001	0.574

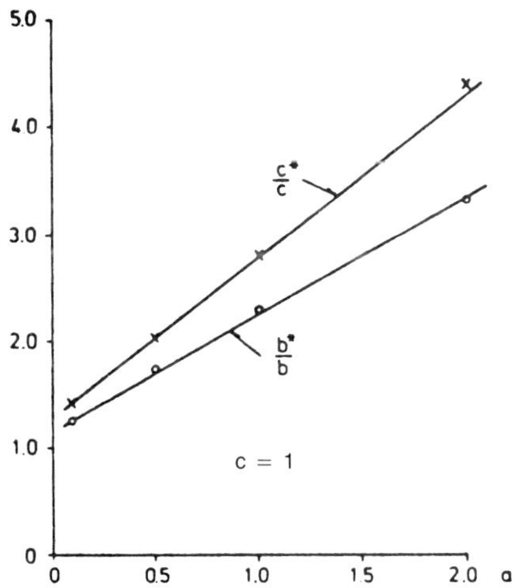

Figure 1. Ratios of the moment estimates to the true parameter values against the location parameter for c = 1.

values of c. Looking at the variance of estimates in Tables 3 and 4, it can be seen that the efficiency of estimating b decreases with increasing value of a for $c = 1$ and remains constant for $c = 10$, while the efficiency of estimating c decreases with increasing value of a for both $c = 1$ and $c = 10$.

In summarizing the Weibull parameter estimation by this method, it may be said that the method cannot be relied on for estimating all three parameters, especially not for higher values of the shape parameter. For two-parameter estimation, the errors in estimation increase linearly with the true value of the location parameter, which is assumed to be zero. The errors are more than

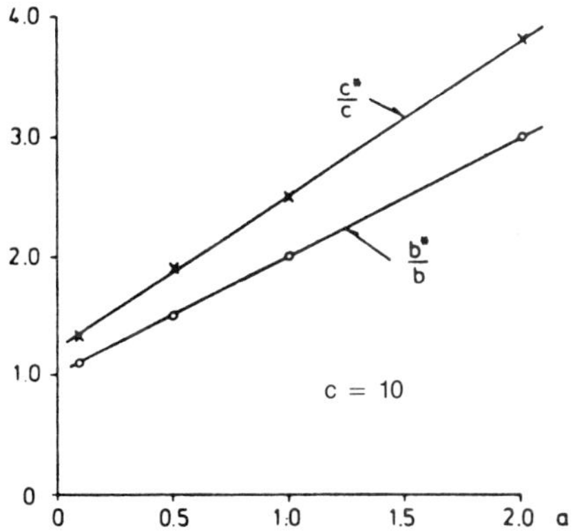

Figure 2. Ratios of the moment estimates to the true parameter values against the location parameter for c = 10.

Table 5. Estimates of Weibull parameters by maximum likelihood estimation assuming a = 0.

Population	(0.1, 1, 1)		(0.5, 1, 1)		(1.0, 1, 1)		(2.0, 1, 1)	
Sample	b*/b	c*/c	b*/b	c*/c	b*/b	c*/c	b*/b	c*/c
1	1.75	1.60	2.21	2.00	2.77	2.50	3.84	3.50
2	0.80	1.40	1.28	2.20	1.82	3.10	2.85	4.80
3	1.22	1.70	1.68	2.50	2.22	3.40	3.25	5.10
4	1.56	1.30	2.08	1.80	2.65	2.40	3.72	3.40
5	1.03	1.30	1.52	1.90	2.07	2.50	3.13	3.70
Average	1.47	1.46	1.75	2.08	2.31	2.78	3.36	4.10
Variance	0.091	0.026	0.120	0.062	0.127	0.158	0.137	0.500

100 percent when the location parameter is equal to or greater than the scale parameter.

MAXIMUM LIKELIHOOD ESTIMATION

No attempt was made here to estimate all three parameters by this method, as the computations involved are tedious and as the estimates may not even be reliable (see section on Maximum Likelihood Estimation).

For two-parameter estimation, the shape parameter is estimated first by maximizing the function in Equation (30) or by solving Equation (31) for c. The former method was used here since it was found to involve simpler computations. After estimating c, b was estimated by using Equation (32) in which the estimate of c was used.

The results of estimation are shown in Table 5 for populations (a, 1, 1) and in Table 6 for populations (a, 1, 10) for $a = 0.1, 0.5, 1.0$, and 2.0. As seen in these tables, the errors in estimating b and c increase with increasing value of the location parameter, which is taken to be zero in the estimation of b and c. The average ratios of the estimates to the true parameter values are plotted

Table 6. Estimates of Weibull parameters by maximum likelihood estimation assuming a = 0.

Population	(0.1, 1, 10)		(0.5, 1, 10)		(1.0, 1, 10)		(2.0, 1, 10)	
Sample	b*/b	c*/c	b*/b	c*/c	b*/b	c*/c	b*/b	c*/c
1	1.15	1.60	1.55	2.10	2.05	2.80	3.05	4.20
2	1.06	1.20	1.46	1.70	1.96	2.20	2.96	3.40
3	1.11	1.60	1.51	2.30	2.01	3.00	3.01	4.50
4	1.13	1.20	1.54	1.70	2.03	2.20	3.04	3.30
5	1.08	1.10	1.49	1.60	1.99	2.10	2.99	3.20
Average	1.11	1.34	1.51	1.88	2.01	2.46	3.01	3.72
Variance	0.001	0.046	0.001	0.074	0.001	0.134	0.001	0.278

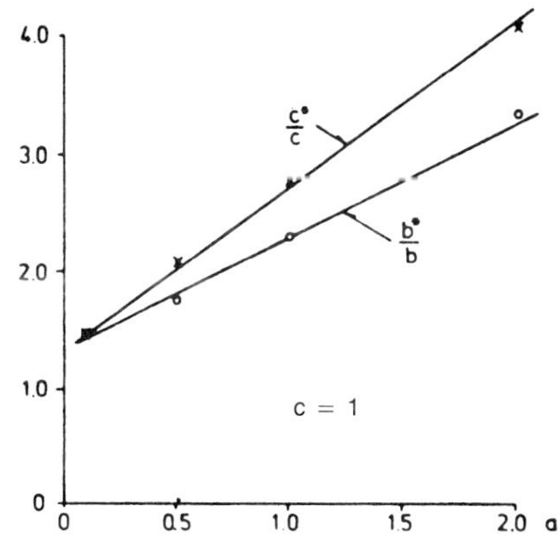

Figure 3. Ratios of the maximum likelihood estimates to the true parameter values against the location parameter for c = 1.

in Figure 3 for $c = 1$ and in Figure 4 for $c = 10$. Both these figures show that the average errors in estimating b and c increase linearly with increasing value of a. By comparing Figures 3 and 4, it is apparent that the estimation errors are lower for higher value of c.

Looking at the variance values in Tables 5 and 6, it is seen that the efficiency of estimating b decreases with increasing value of a for $c = 1$ and remains constant for $c = 10$. This constant variance value is the same as that obtained in estimating b for $c = 10$ by the moment estimation method, as seen in Table 4. The efficiency in estimating c decreases with increasing value of a both for $c = 1$ and $c = 10$.

The estimates of c can be improved slightly by multiplying these by the unbiasing factor, which, for $N = 10$, is given by Thoman, Bain, and Antle [6] as 0.859.

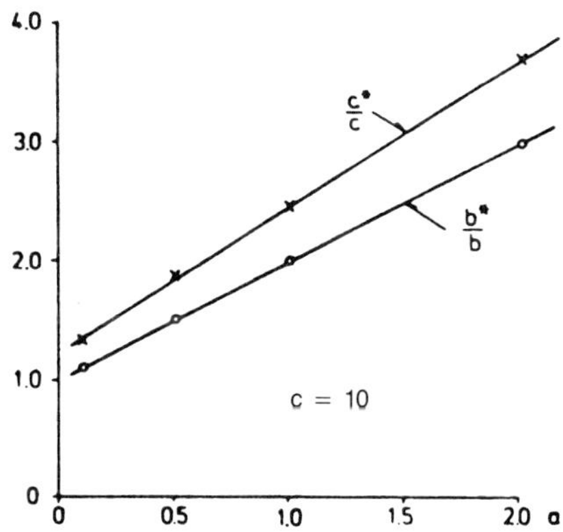

Figure 4. Ratios of the maximum likelihood estimates to the true parameter values against the location parameter for c = 10.

Table 7. The expected value, median value, and variance of the order statistic z_i for a sample of size 10.

i	Ez_i, c = 1	Mz_i, c = 1	Vz_i, c = 1	Vz_i, c = 10
1	0.10000	0.06905	0.01000	0.00827
2	0.21111	0.17700	0.02235	0.00426
3	0.33611	0.29916	0.03797	0.00298
4	0.47897	0.43864	0.05838	0.00236
5	0.64564	0.60092	0.08616	0.00200
6	0.84564	0.79476	0.12616	0.00178
7	1.09564	1.03539	0.18866	0.00164
8	1.42897	1.35264	0.29977	0.00159
9	1.92897	1.81878	0.54977	0.00166
10	2.92897	2.70725	1.54977	0.00209

STANDARDIZED VARIABLE ESTIMATION

To start estimation by this method, one must plot the order statistics, x_i, against Ez_i or Mz_i for a trial value of c. It is advisable to start with $c = 1$. In Table 7, values of Ez_i and Mz_i are listed for $c = 1$ for a sample of size 10. For other values of c and sample sizes, see Weibull [9] and Weibull and Weibull [10].

To illustrate the method, we first take a random sample from a Weibull population (0.1, 1, 1). The (x_i, Mz_i) plots are shown for $c = 0.5$, 1.0, and 2.0 in Figure 5. The consecutive points are joined by a straight line. As seen here, the plot curves to the right for $c = 0.5$, which is less than the true value of c. For the true value of c(that is, $c = 1$), the data appear to scatter about a straight line. For a higher value of c, the scatter appears to increase. Thus, an inspection of the (x_i, Mz_i) plots gives an idea of the range in which a search for the estimate of c should be made. Once this is done, an estimate of c is found by making a linear regression analysis of the (x_i, Mz_i) data points for different values of c and choosing as estimate the value that maximizes the correlation coefficient. For the data plotted in Figure 5, the following correlation coefficients r were obtained.

$c = 1.0$	1.1	1.2	1.3
$r = 0.99359$	0.99527	0.99561	0.99509

Thus, the estimate $c^* = 1.2$ is taken. For further accuracy, interpolation between the c-values may be made. The estimates of a and b are now given by the x_i-intercept and the slope of the best fit line.

Table 8 shows the estimates of all three parameters for five samples from Weibull population (0, 1, 1). For Samples 3 and 4, small negative estimates of a are obtained. To avoid these unacceptable estimates, we could look for

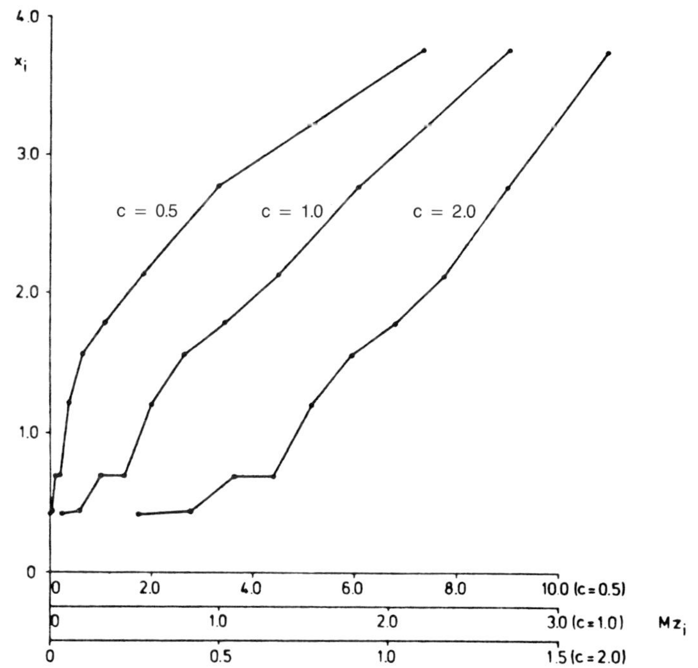

Figure 5. The (x_i, Mz_i) plots for a random sample from Weibull population (0.1, 1, 1).

the c-values that maximize the correlation coefficient within the constraint $a \geq 0$. Doing this, revised estimates for Samples 3 and 4 are obtained as (0, 1.22, 1.4) and (0, 1.46, 1.1), respectively. These estimates are indeed better than the corresponding estimates in Table 8.

If the estimates of a given by this method are positive, then the estimates of b and c will remain unchanged for higher values of a. However, if a is estimated to be negative, then the constraint $a \geq 0$ for estimation would change (improve) the estimates of b and c for higher values of a. This is illustrated

Table 8. Estimates of Weibull parameters by standardized variable estimation.

Population	(0, 1, 1)		
Sample	a*	b*/b	c*/c
1	0.02	1.60	1.20
2	0.0	0.68	1.00
3	−0.37	1.55	2.00
4	−0.20	1.74	1.30
5	0.01	0.86	0.90
Average	−0.11	1.29	1.28
Variance	0.024	0.186	0.150

for a set of samples taken from Weibull populations (a, 1, 10) with a = 0.1, 0.5, 1.0, and 2.0 below.

The (x_i, Mz_i) plots for one sample from population (0.1, 1, 10) are shown in Figure 6 for c = 1 and c = 10. For c = 1, the plot curves to the right indicating that a higher value of c should be taken. For c = 10, the data points, except the lowest one, scatter about a straight line. This behavior of the lowest data point is typical for high values of c and is explained by the variance of the lowest order statistic (see Table 7). As seen in this table, Vz_i for c = 10 is much higher than the remaining Vz_i. It is therefore advisable to censor the first order statistic when the (x_i, Mz_i) plots indicate a high value of c. This was done for all samples from Weibull populations (a, 1, 10). Furthermore, the constraint $a \geq 0$ was employed in estimation as a^* otherwise became negative. The estimates are shown in Table 9.

As seen in Table 9, accurate estimates of b and c are obtained at the lowest value of a, and the accuracy diminishes at higher values of a. Beyond a = 1, however, the estimation errors increase at a lower rate WRT a. The efficiency of estimating both b and c decreases with increasing value of a.

In summarizing the standardized variable estimation method, it may be said that this provides accurate and efficient estimates of all three Weibull parameters for low values of c. For high values of c, however, it tends to give negative estimates of the location parameter, and on applying the constraint $a \geq 0$, it estimates b and c with better accuracy and efficiency than the other methods just described.

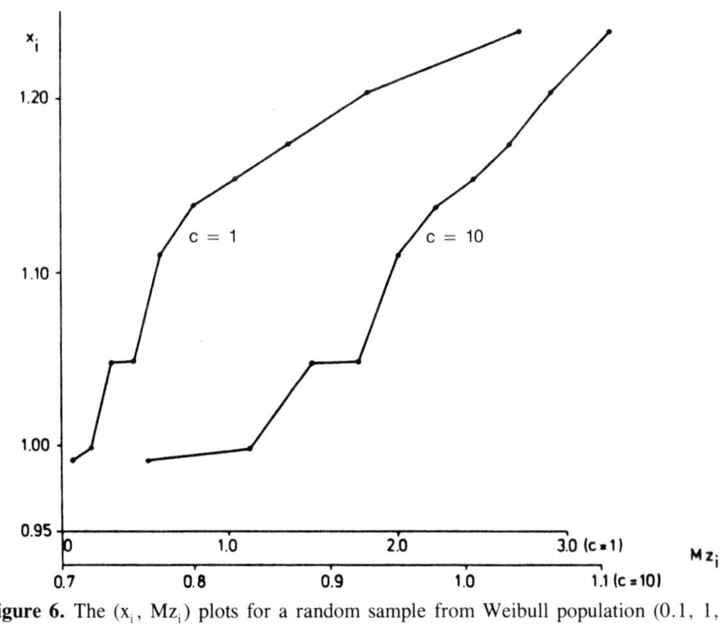

Figure 6. The (x_i, Mz_i) plots for a random sample from Weibull population (0.1, 1, 10).

Table 9. Estimates of Weibull parameters by standardized variable estimation with the condition a ≥ 0.

Population	(0.1, 1, 10)		(0.5, 1, 10)		(1.0, 1, 10)		(2.0, 1, 10)	
Sample	b*/b	c*/c	b*/b	c*/c	b*/b	c*/c	b*/b	c*/c
1	1.11	1.20	1.54	1.84	2.01	2.20	2.19	2.40
2	1.08	1.00	1.46	1.35	1.95	1.80	2.16	2.00
3	1.13	1.30	1.56	1.80	2.07	2.40	3.00	3.5
4	1.13	1.20	1.50	1.60	1.68	1.80	1.68	1.80
5	1.10	1.00	1.52	1.40	2.02	1.90	2.95	2.90
Average	1.11	1.14	1.52	1.60	1.95	2.02	2.40	2.52
Variance	0.0004	0.014	0.001	0.040	0.019	0.058	0.256	0.382

Estimation from Composite Material Strength and Fatigue Life Data

To illustrate the use of the standardized variable estimation method further, we take as examples some recently published data on composite material strength and fatigue life.

As the first example, we take the compression test results for two graphite/epoxy laminates reported by Ryder and Black [11]. Their test results, given in their Table 1 (Laminate 1) and Table 2 (Laminate 2), are plotted as (x_i, Mz_i) plots for a trial value of $c = 1$ in Figures 7 and 8. These two figures show that the data points scatter about two straight lines in each

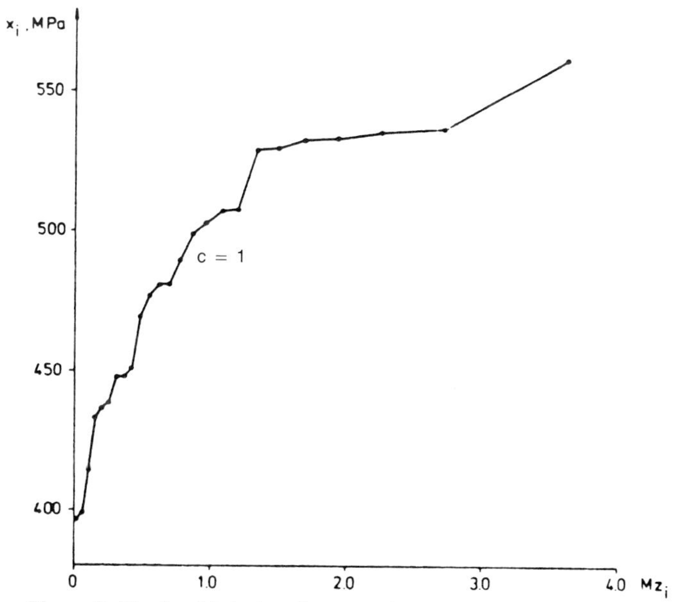

Figure 7. The (x_i, Mz_i) plot of compression test data for Laminate 1.

plot. The highest point in each case, however, appears to be an outlier. It is therefore apparent that, except for the highest point, the data belong to a two-component population. According to the authors, the stress-strain plots indicated that two different modes of fracture existed. The (x_i, Mz_i) plots therefore appear to confirm this observation. The Laminate 1 data in Figure 7 indicate that the seven highest points (except perhaps the highest point) belong to one component and the remainder to another. According to the authors' observation of the stress-strain plots, however, there should be eleven highest points in a component. A linear regression analysis of the data points in the lower component showed that inclusion of the four highest points increased the correlation coefficient. We therefore tend to think that only the seven highest points belong to a higher component. The data for Laminate 2 plotted in Figure 8 confirm the authors' observation that the nine highest data points belong to Mode 2 fracture.

The estimation procedure described in a previous section was applied to each component separately, and the estimates so obtained are listed below.

Laminate 1:
Component 1: $a^* = 365.8$ MPa, $b^* = 138.0$ MPa, $c^* = 2.5$
Component 2: $a^* = 422.1$ MPa, $b^* = 104.4$ MPa, $c^* = 10.0$
Laminate 2:
Component 1: $a^* = 713.0$ MPa, $b^* = 109.4$ MPa, $c^* = 12.5$
Component 2: $a^* = 739.8$ MPa, $b^* = 71.1$ MPa, $c^* = 7.0$

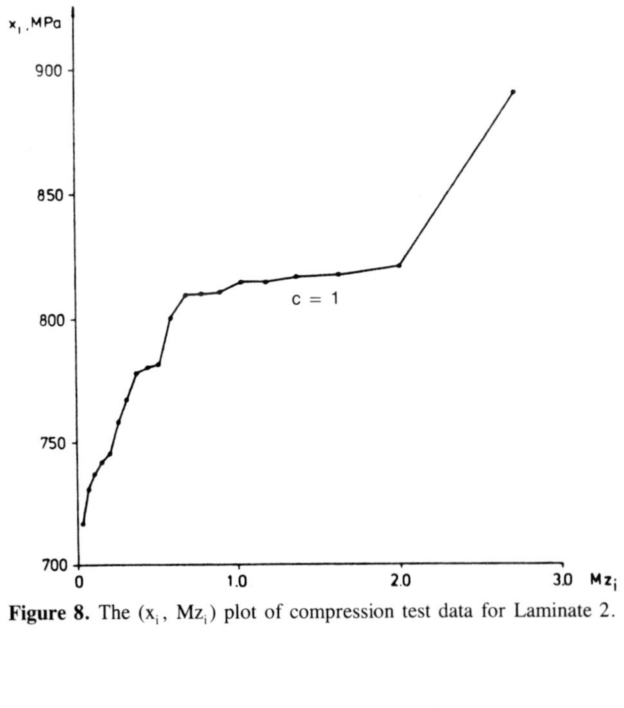

Figure 8. The (x_i, Mz_i) plot of compression test data for Laminate 2.

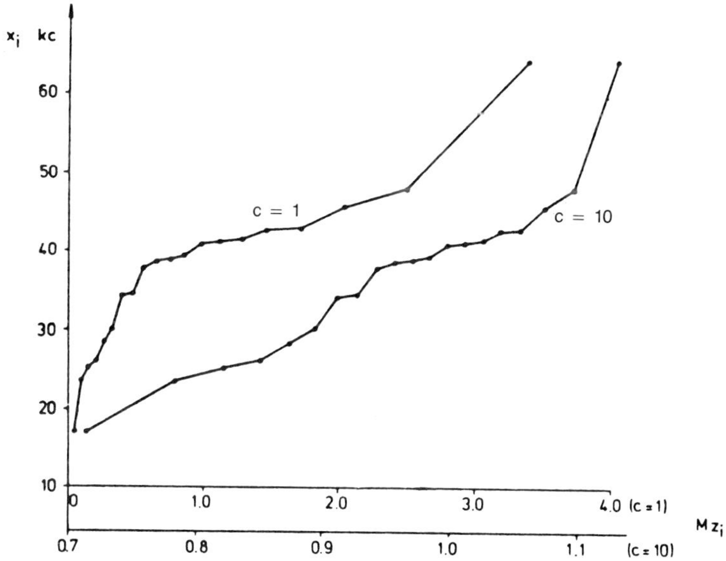

Figure 9. The (x_i, Mz_i) plots of fatigue test data.

The second example taken is the fatigue life data for graphite/epoxy $[\pm 45]_{2s}$ laminate tested in shear at maximum stress of 8.143 ksi and $R = 0.1$ reported by Yang and Jones [12]. These data, consisting of 20 points, are plotted as x_i in kilocycles against Mz_i for trial values of $c = 1$ and $c = 10$ in Figure 9. It is clearly seen that, except for the highest point, the data are homogeneous and have a shape parameter higher than 1, since the (x_i, Mz_i) plot for $c = 1$ curves to the right. The plot for $c = 10$ shows no more curvature to the right and confirms that the highest point is an outlier. We therefore censor this observation and estimate the parameters from the remaining data.

The estimation procedure gave a negative value of a, whereupon the constraint $a \geq 0$ was placed and the following estimates were obtained.

$$a^* = 0, \quad b^* = 38.14\text{kc}, \quad c^* = 4.21$$

These estimates may be compared with Yang and Jones' predictions of the two parameters (assuming $a = 0$), which were calculated to be

$$b^* = 41.07\text{kc}, \quad c^* = 4.31$$

Conclusions

It has been demonstrated that the moment estimation and maximum likelihood estimation methods lead to large errors in estimating the scale and the shape parameters, if the location parameter is taken to be zero. The errors increase linearly with the true value of the location parameter.

The proposed method, called here the standardized variable estimation, has been demonstrated to give accurate and efficient estimates of all three Weibull parameters for low shape parameters, and more accurate and more efficient estimates than the other two methods of the scale and the shape parameters for high shape parameter. Furthermore, this method allows separation of components in a multi-component sample and detection of outliers.

Acknowledgements

The author would like to express his grateful appreciation of the inspiration and guidance he received from collaboration with Professor W. Weibull, who died recently having contributed actively to the field of statistical strength theories until his very last days.

References

1. Weibull, W. and Weibull, G. W., "New Aspects and Methods of Statistical Analysis of Test Data with Special Reference to the Normal, the Lognormal, and the Weibull Distributions," Part I and Part II, *FOA Report D 20045-DB*, Defence Research Institute, Stockholm (June 1977).
2. Gumbel, E. J., *Statistics of Extremes*, Columbia University Press, N.Y. (1960).
3. Antle, C. E. and Klimko, L. A., "Choice of Model for Reliability Studies and Related Topics II," ARL-73-0121, AD 772775 (1973).
4. Harter, H. L. and Moore, A. H., *Technometrics*, 7, No. 4 pp. 639–643 (1965).
5. Cohen, A. C., *Technometrics*, 7, No. 4, pp. 579–588 (1965).
6. Thoman, D. R., Bain, L. J. and Antle, C. E., "Inferences on the Parameters of the Weibull Distribution," *Technometrics*, Vol. 11, No. 3 (1969).
7. McCool, J. I., in *Transactions*, 13, American Society of Lubrication Engineers, pp. 189–202 (1969).
8. Lieblein, J., *Annals of Mathematical Statistics*, 26, pp. 330–333 (1955).
9. Weibull, W., "Estimation of Distribution Parameters by a Combination of the Best Linear Order Statistics and Maximum Likelihood," AFML-TR-67-105, Air Force Materials Laboratory (1967).
10. Weibull, W. and Weibull, G. W., "High-fidelity Approximation to Median Percentage Points of Order Statistics," *AFML-TR-69-317*, Air Force Materials Laboratory (1969).
11. Ryder, J. T. and Black, E. D. in *Composite Materials: Testing and Design (Fourth Conference)*, *ASTM STP 617*, American Society for Testing and Materials, pp. 170–189 (1977).
12. Yang, J. N. and Jones, D. L., *Journal of Composite Materials*, 13, pp. 371–389 (1978).

CHAPTER 10

On Design Criteria for Composite Structures Under Static and Fatigue Loads

Abstract: Design criteria for composite structures are discussed in relation to the critical events in the development of damage in composite laminates under static loads and under fatigue loads. For static loads, the currently used criteria, namely, the first ply failure criterion and the strength criterion are evaluated and the use of stiffness degradation criterion is discussed. For fatigue loads, a two-stage strength degradation model is proposed for reliability analysis.
Keywords: Composite structural design, composite materials damage, composite stiffness degradation, composite strength degradation, composite reliability.

ADVANCED COMPOSITE MATERIALS, e.g. graphite-epoxy are gaining applications as structural materials at an increasing rate. This has given rise to an urgent need for developing analyses for assessment of safety and reliability of composite structures. Most studies hitherto made toward this end have been inspired by experience with metallic structures. For instance, the approach to fatigue reliability of composite laminates based on strength degradation models [1,2] is basically that used for metallic structures [3–5] with the exception that the strength degradation is specifically related to crack growth in metals, while it is assumed to result from some damage growth process in composites. For increased confidence in such analyses, however, it is desirable to clarify the underlying damage mechanisms and to consider the likely causes of statistical behaviour.

We shall here discuss the evolution of damage in composite laminates under mechanical loads and consider the critical events in the damage process that might be taken as bases for structural design. For the case of static loading, we shall evaluate the design criteria presently in use, namely, the first ply failure criterion and the strength (total failure) criterion, and discuss stiffness

Reprinted by permission from *Structural Safety and Reliability*, Vol. 1, Proceedings of ICOSSAR '85, Kobe, Japan, pp. 425–434 (May 1985).

degradation as a possible design criterion. For fatigue loading, we shall propose a two-stage strength degradation model for reliability analysis on the basis of the observed damage development in composite laminates.

Static Loads

MATERIALS RESPONSE

Consider a composite laminate consisting of a number of layers (plies) with different fiber orientations. On loading, the first significant damage event is the appearance of matrix cracks in a ply (or plies) having fibers along a direction closest to the normal to the principal tensile load. This occurs when the stress (or strain) normal to fibers in the ply (or plies) exceeds the threshold value for onset of cracking. On further increase of loading, more cracks occur parallel to the previously initiated cracks and, when the cracking threshold for other plies of less favorably oriented fibers is exceeded, these plies crack too. The crack initiation and multiplication process continues with loading until either a saturation state is reached, where cracks in individual plies have achieved stable spacings, or fibers in a ply (or plies) closest to the principal tensile stress direction fail.

Consider now a laminate with two sets of plies: one set which undergoes cracking and the other set which applies constraint to this cracking, i.e. it restricts the crack surface displacements. We shall call the two sets as the cracking plies and the constraining plies, respectively. Figure 1 shows an edge view of a laminate illustrating the two ways of mixing the two sets of plies. The size of an individual crack is limited by the total thickness of the adjacent cracking plies between the constraining plies. The effectiveness of the constraint to the cracking process depends on the size of crack and on the stiffness of the constraining plies. Figure 2 illustrates the constraint by showing the variation of the applied tensile strain at which cracking initiates with the number of adjacent cracking plies (which is proportional to the crack size) for a given set of constraining plies (which, in this case, consists of four longitudinal plies on either side of the transverse cracking plies). Similar effect is seen when the crack size is fixed and the stiffness of the constraining plies is varied. The degree of constraint may be characterized by four types, as indicated in Figure 2, namely, no constraint (A), low constraint (B), high constraint (C) and full constraint (D).

Figures 3–6 show the stress-strain response corresponding to the four types of constraint. The limiting case of no constraint (A) exhibits behaviour which is seemingly identical to the elastic-perfect plastic behaviour in metals, although the underlying mechanisms are quite different in the two cases. When the applied load reaches the threshold value for crack initiation, denoted by ϵ_{FPF}, i.e., the first ply failure strain, the laminate deforms unin-

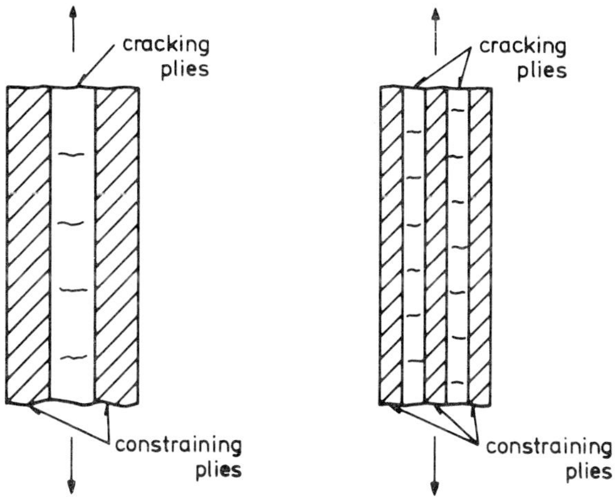

Figure 1. Edge view of laminates undergoing cracking: a) cracking plies are lumped together, b) cracking plies are distributed.

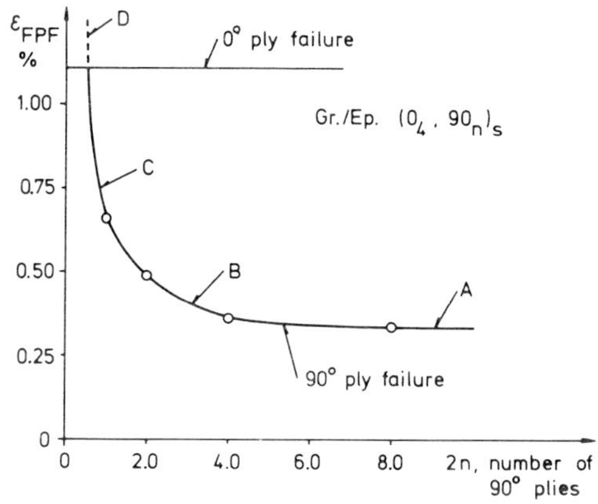

Figure 2. Effect of constraint on the strain to onset of matrix cracking, ϵ_{FPF}.

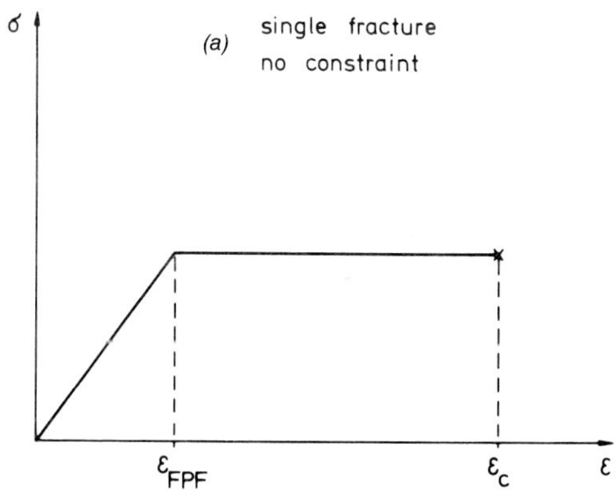

Figure 3. Schematic stress-strain response for type A constraint.

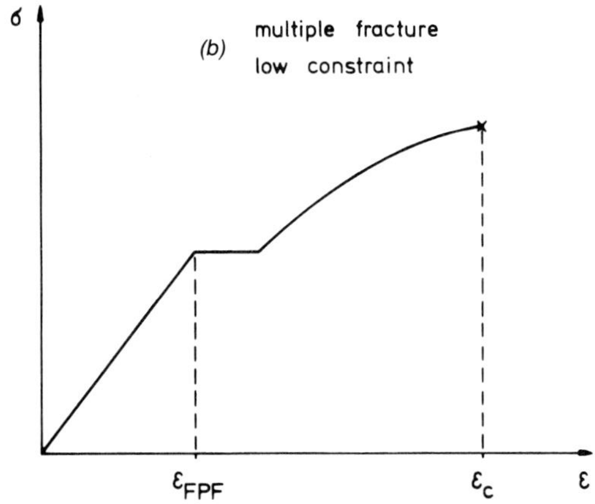

Figure 4. Schematic stress-strain response for type B constraint.

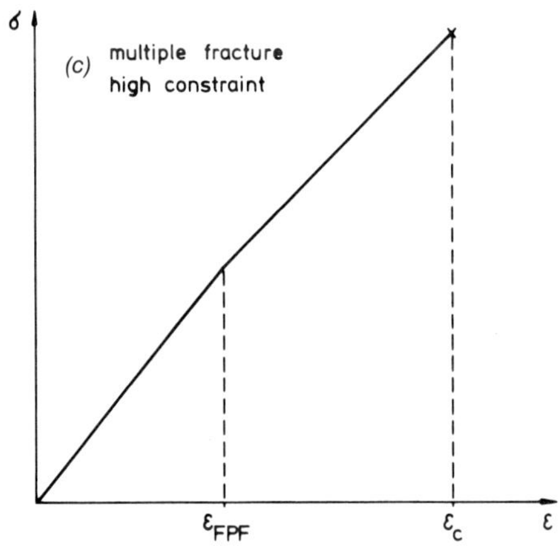

Figure 5. Schematic stress-strain response for type C constraint.

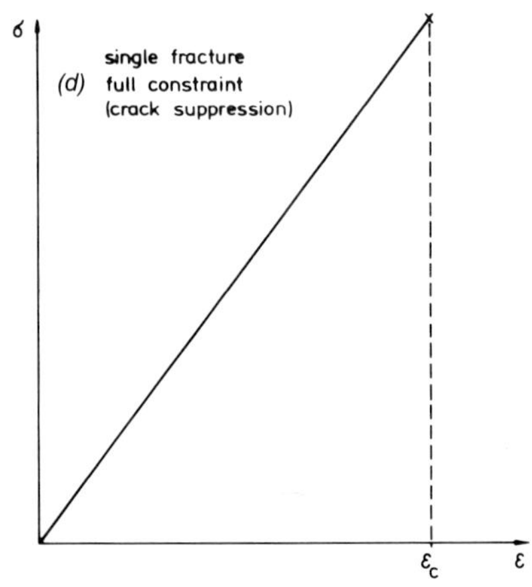

Figure 6. Schematic stress-strain response for type D constraint.

hibitedly until the strain reaches the failure strain for the constraining plies. The threshold strain, ϵ_{FPF}, increases with the constraint, and the post cracking behaviour is affected increasingly with the constraint, as illustrated by Figures 4 and 5. In Figure 6 the limiting case of full constraint is illustrated. The initiation of cracking is fully suppressed here, i.e. the constraining plies fail first.

The description of damage in composite laminates and the consequent mechanical response, cited here briefly, has been extracted from Chapter 8, wherein further references can also be found. We shall use this information to discuss the design criteria under static loads in the following.

DESIGN CRITERIA

A design criterion for structures under mechanical loads must account for a critical (and often undesirable) event in the structural response. Examples of critical events under static loads are yielding, fracture and instability, while the parameters representing these events are yield stress, fracture toughness, critical crack size, buckling stress, etc.

For composite laminates under static loads, the materials response exhibits initiation of matrix cracking in one or more plies as the first critical event. The parameter representing this event is the threshold stress (or strain), denoted by σ_{FPF} (or ϵ_{FPF}). We have illustrated the effect of cracking constraint on this parameter in the discussion above. The last critical event is fracture (separation) of laminate which results from fiber failures in plies most closely aligned with the principal tensile stress. The parameter representing this event is strength, denoted by σ_c (or ϵ_c).

The relative occurrence of the two critical events is illustrated in Figure 2 on the strain scale. It is clear from the figure that, at low constraint, the laminate can sustain an appreciable load beyond the first ply failure (cracking), while in the extreme case of full constraint, the laminate fails before any matrix cracking. This illustrates the necessity of considering the laminate configuration and, in particular the crack constraining properties, before selecting a design criterion.

The criticality of an event in the structural response depends on the structural function. The event of total fracture (separation) would be critical in all cases. However, the extent of first ply failure (matrix cracking) may not always be critical. If the structural function is to carry loads, then, as we have seen above, appreciable loads can be carried beyond the first ply failure. However, the extent of deformation may become critical (or untolerable) before the load bearing capacity is exhausted. The deformational response of a structure is determined by its stiffness and it is this property we shall focus on in further discussion of the design criteria.

STIFFNESS DEGRADATION AS A DESIGN CRITERION

The stress-strain response of a laminate undergoing cracking is illustrated in Figure 7. The initial linear elastic behaviour, represented by the modulus E_1^0, discontinues when the threshold strain, ϵ_{FPF}, is reached, and the post cracking response is, in general, nonlinear and is determined by the cracking constraints. If the laminate is loaded to a point beyond the cracking threshold and unloaded, then, cracks close as the stresses are relieved and at zero load, the laminate is unstrained (assuming no curing stresses). On reloading the laminate behaves as a linear elastic material following the path with slope E_1 (shown dotted in the figure) until the point of previous unloading is reached. The current elastic modulus thus reflects the current state of damage, and the stiffness degradation, $(E_1^0 - E_1)$, represents the damage sustained by the laminate. This forms the basis for formulating a design criterion in terms of stiffness degradation.

The relationship between stiffness degradation and the extent of damage in composites has been studied by various authors [7–10] and in chapter 6 above. A fairly general form of the stiffness-damage relationships was derived by the present author on the basis of a continuum mechanics characterization in Chapter 6. As an example, the relationships expressing changes in the engineering elastic moduli for cracks of one orientation in a thin orthotropic laminate are given below:

$$E_1 = E_1^0 - 2D^2[k_3 + k_7(\nu_{12}^0)^2 - k_{13}\nu_{12}^0]$$

$$E_2 = E_2^0 - 2D^2[k_7 + k_3(\nu^0{}_{21})^2 - k_{13}\nu_{12}^0]$$

$$\nu_{12} = \nu_{12}^0 - D^2\left[\frac{1 - \nu_{12}^0\nu_{21}^0}{E_2^0}\right](k_{13} - 2k_7\nu_{12}^0)$$

and

$$G_{12} = G_{12}^0 - 2D^2k_{11} \tag{1}$$

where

$$D^2 = \eta_c \cdot \bar{\ell}_c \cdot \bar{w}_c \cdot f_c \tag{2}$$

where η_c is the number of cracks per unit surface area and $\bar{\ell}_c$ and $\bar{w}_c$ are the average length and width, respectively, of the cracks. The factor f_c accounts for the constraint to cracking, and is taken, in a simple form, as the ratio of the thickness of the cracking plies to the thickness of the laminate. The constants k_i are material constants, and the elastic moduli are written in the usual notation.

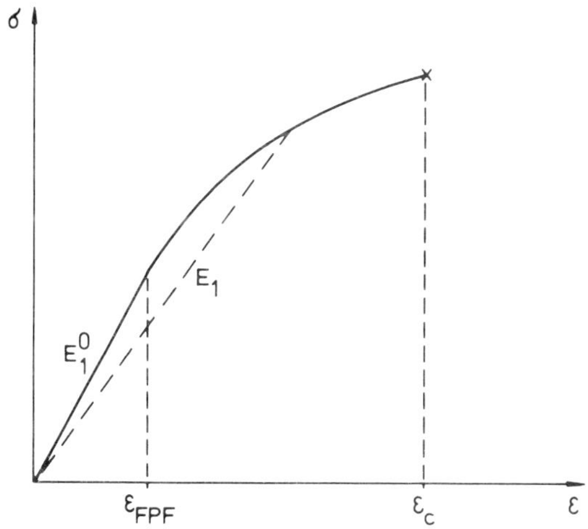

Figure 7. Stress-strain response of a laminate undergoing cracking.

For a given laminate, the damage parameter, D^2, depends on the crack number density, η_c, which in turn depends on the applied load. Thus, having a relationship between crack density and load, determined analytically or emperically, we can calculate degradation of a stiffness property for a given load by using Equations (1) and (2). A basis for determining design load for a critical stiffness degradation is thus established.

A NOTE ON RELIABILITY ANALYSIS

For structural design based on stiffness degradation, the probability of failure may be defined as the probability of stiffness degradation to a critical value. The scatter in the residual elastic moduli results from the statistical nature of the cracking process, which is expressed in the randomness of the crack number density in Equation (2). A reliability approach would now be to formulate a statistical model of the cracking process, which could provide probability distribution of the crack density parameter. Using Equations (1) and (2), then, the probability distributions of the residual elastic moduli could be determined.

Fatigue Loads

FATIGUE DAMAGE

Fatigue damage in composite materials has been studied by several researchers, see for example, References [12–14] and Chapters 2 and 4. We shall

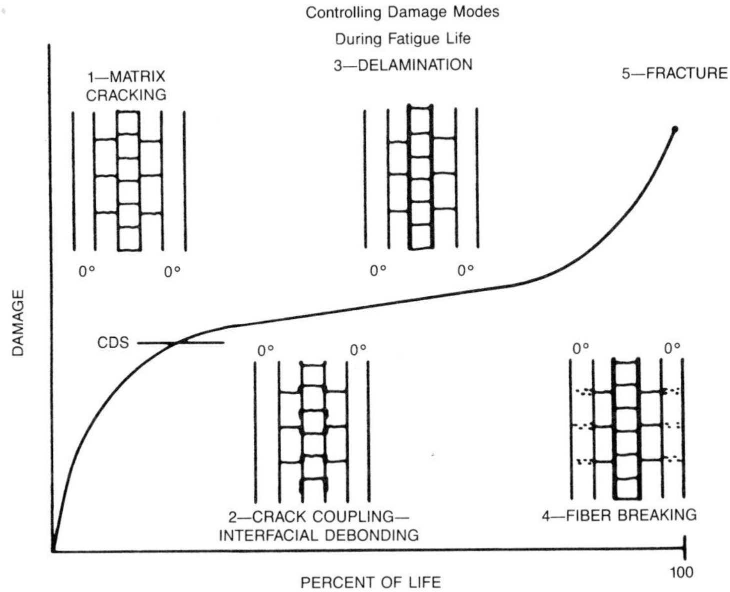

Figure 8. Development of damage in composite laminates under fatigue (From [14]).

here give a brief description of the evolution of damage, extracted from these references.

Figure 8 shows a schematic view of the development of damage under tensile fatigue for a large class of composite laminates. In the initial stage, matrix cracks form along fibers in off-axis plies, i.e. plies with fiber axis not aligned with the principal tensile stress direction. These cracks increase in number with load cycles, and eventually a saturation state ensues, wherein a stable crack pattern develops. This state has been called a Characteristic Damage State, CDS, and has been found to be independent of the loading history, but determined by the laminate properties, i.e. ply stiffnesses and ply stacking sequence. As the matrix cracks form, fiber failures initiate preferentially in adjacent plies in regions of stress concentration created by the cracks. The matrix cracks, which form entirely through the thickness of the off-axis plies and generally run across the total width of a uniformly stressed region, also initiate short cracks in adjacent plies. At the ply interfaces, close to the intersection of the long and the short cracks, strong interlaminar stresses develop which lead to separation of the plies locally. The later stage of damage is typified by a rapidly increasing rate of progression of all damage modes in a rapidly decreasing volume culminating in a critical local state of stress which initiates fracture of the laminate.

The damage process may be considered as consisting of two dominant stages: the first stage in which noninteracting cracks develop leading to attainment of a characteristic damage state, and the second stage in which cracks of various types and orientations interact with increasing rates causing increasing localization and consequent failure.

FATIGUE RELIABILITY

Before proceeding with a reliability analysis of composites under fatigue, we note the following properties of the fatigue damage.

1. The damage evolves in two stages. In the first stage, the damage is homogeneous and noninteractive. In the second stage, the damage tends to localize in zones of increasing interaction between the damage modes.
2. The transition from the first stage to the second stage occurs at CDS, the characteristic damage stage, which consists of a well-defined crack pattern characterizing the saturation of the noninteractive damage.

Based on these properties, we propose a two-stage strength degradation model.

STAGE I

In this stage a general weakening of the material takes place as a result of cracking in the various plies. Let us now define a damage parameter, D, in terms of this weakening as

$$D = \frac{R_0 - R}{R_0 - R_c} \tag{3}$$

where R_0 is the initial strength, R is the current (residual) strength, and R_c is the strength of CDS.

We assume further that the rate of fatigue damage depends on the "effective stress," and is given by

$$\frac{dD}{dN} = k\left(\frac{S}{1 - D}\right)^m \tag{4}$$

where S is the maximum stress and k and m are material constants.

Integrating Equation (4) and using Equation (3), we obtain the relationship between the initial strength and the residual strength as

$$R = R_c + (R_0 - R_c)(1 - N/N_c)^{m'} \tag{5}$$

where

$$m' = \frac{1}{1 + m} \text{ and } N_c = \frac{1}{k(1 + m)S^m}$$

Equation (5) may be used to obtain the probability distribution of R when the probability distributions of the variables on the right hand side of the equation are known. The initial strength, R_0, is a random variable due to the statistical variation of the fiber strength, and its probability distribution may be determined experimentally. The variation in R_c is expected to be negligible since it is the strength of a composite at CDS, which is deterministically given by a mechanics analysis, e.g. a shear lag analysis [8]. N_c is the number of cycles to attainment of CDS, and it would be random due to the randomness of the rate of crack density increase, assuming that the cracks initiate from flaws which are randomly distributed in the material volume. We shall discuss the determination of the probability distribution of N_c in the following. At this point it may be noted that having the distributions of R_0 and N_c, the distribution of R, and hence the probability of failure given by $P(R \leq S)$, can be determined numerically by using Equation (5).

STAGE II

In this stage the damage is localized in a few zones where the intensity of crack interactions is high and increasing with fatigue cycling. These zones may be conceptually substituted by a single crack which releases the same amount of the stored elastic energy as that released combinely by the various damage mechanisms in all the localized zones. Assuming that this crack has a characteristic dimension given by C, we may express the residual strength as

$$R = \alpha C^{-1/2} \tag{6}$$

where α is a material constant characterizing the material toughness.

The rate of damage growth may now be expressed by the rate of increase of the crack size C, and taken as

$$\frac{dC}{dN} = \beta C^{n/2} \tag{7}$$

where β is a function of stress.

From Equations (6) and (7), we obtain,

$$\frac{dR}{dN} = -\gamma \left(\frac{1}{R} \right)^{n-3} \tag{8}$$

where $\gamma = \beta/2(\alpha)^{n-2}$.

Assuming $n > 2$, and integrating Equation (8) between N and N_f, the number of cycles to failure, we obtain,

$$R^{n-2} = S^{n-2} + \gamma(n-2)(N_f - N) \tag{9}$$

This equation provides a means of determining the probability distribution of R^{n-2} when the probability distribution of N_f is given. Thus, let N_f, the fatigue life, have a Weibull distribution, then, we have,

$$P(R^{n-2} \leq r^{n-2}) = 1 - \exp\left[-\left(\frac{r'' - a}{b}\right)^{c}\right] \tag{10}$$

where $r'' = r^{n-2}$, $a = S^{n-2} + \gamma(n-2)(a_n - N)$, $b = \gamma(n-2)b_n$, and $c = c_n$ where a_n, b_n and c_n are the location, scale and shape parameters, respectively, of the fatigue life distribution.

Taking $R = R_c$ for $N = N_c$ in Equation (9), we have,

$$N_c = N_f - \frac{R_c^{n-2} - S^{n-2}}{\gamma(n-2)} \tag{11}$$

The probability distribution of N_c is now given by

$$P(N_c \leq N) = 1 - \exp\left[-\left(\frac{N - a_n'}{b_n}\right)^{c_n}\right] \tag{12}$$

where

$$a_n' = a_n - \frac{R_c^{n-2} - S^{n-2}}{\gamma(n-2)}$$

Figure 9 shows the two-stage strength degradation model for fatigue reliability of composite materials.

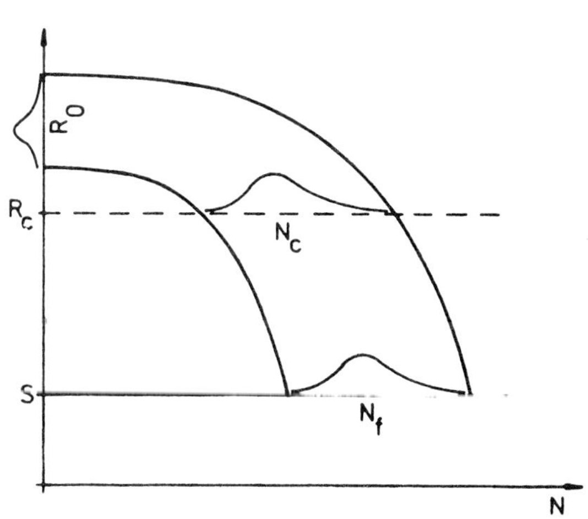

Figure 9. Two-stage strength degradation model for fatigue reliability of composites.

References

1. Hahn, H. T. and Kim, R. Y., "Proof Testing of Composite Materials," *Journal of Composite Materials*, 9, pp. 297–311 (July 1975).
2. Yang, J. N. and Liu, M. D., "Residual Strength Degradation Model and Theory of Periodic Proof Tests for Graphite/Epoxy Laminates," *Journal of Composite Materials*, 11, pp. 176–203 (April 1977).
3. Yang, J. N. and Trapp, W. J., "Reliability Analysis of Aircraft Structures Under Random Loading and Periodic Inspection," *AIAA Journal*, 12, No. 12, pp. 1623–1630 (1974).
4. Talreja, R. and Weibull, W., "Probability of Fatigue Failure Based on Residual Strength," *Fracture 1977*, Proc. ICF4, Pergamon Press, pp. 1125–1131 (1977).
5. Talreja, R., "Fatigue Reliability Under Multiple Amplitude Loads," *Engineering Fracture Mechanics*, 11, pp. 839–849 (1979).
6. Talreja, R., "Transverse Cracking and Stiffness Reduction in Composite Laminates," *DCAMM Report No. 294*, Danish Center for Applied Mathematics and Mechanics, the Technical University of Denmark, Lyngby, Chapter 8 in the present volume (Sept. 1984).
7. Hahn, H. T. and Tsai, S. W., "On the Behavior of Composite Laminates after Initial Failures," *Journal of Composite Materials*, 8, pp. 288–305 (1974).
8. Highsmith, A. L. and Reifsnider, K. L., "Stiffness-Reduction Mechanisms in Composite Laminates," *Damage in Composite Materials*, ASTM STP 775, American Society for Testing and Materials, Philadelphia, pp. 103–117 (1982).
9. Gottesman, T., Hashin, Z. and Brull, M. A., "Effective Elastic Moduli of Cracked Fiber Composites," *Advances in Composite Materials*, Proc. ICCM3, Pergamon Press, pp. 749–758 (1980).
10. Laws, N., Dvorak, G. J. and Hejazi, M., "Stiffness Changes in Unidirectional Composites Caused by Crack Systems," *Mechanics of Materials*, 2, pp. 123–137 (1983).
11. Talreja, R., "A Continuum Mechanics Characterization of Damage in Composite Materials," to appear in *Proc. Roy. Soc. London*, Chapter 6 in the present volume.
12. Hahn, H. T., "Fatigue Behavior and Life Prediction of Composite Laminates," *Composite Materials: Testing and Design* (Fifth Conference), ASTM STP 674, American Society for Testing and Materials, Philadelphia, pp. 383–417 (1979).
13. Stinchcomb, W. W. and Reifsnider, K. L., "Fatigue Damage Mechanisms in Composite Materials," *Fatigue Mechanisms*, ASTM STP 675, American Society for Testing and Materials, Philadelphia, pp. 762–787 (1975).
14. Reifsnider, K. L., Henneke, E. G., Stinchcomb, W. W. and Duke, J. C., "Damage Mechanics and NDE of Composite Laminates," *Mechanics of Composite Materials*, Pergamon Press, pp. 399–420 (1983).
15. Talreja, R., "Fatigue of Composite Materials: Damage Mechanisms and Fatigue Life Diagrams," *Proc. Roy. Soc. London*, A378, pp. 461–475, Chapter 2 in the present volume (1981).
16. Talreja, R., "A Conceptual Framework for the Interpretation of Fatigue Damage Mechanisms in Composite Materials," to appear in *Composite Technology Review*, Chapter 4 in the present volume.

Index